Sustainable Transitions in India
Progress and Directions

Sustainable Transitions in India Progress and Directions

Editors

J.V. Arun
A. Premkumar

Published by

Kalpaz Publications
C-30, Satyawati Nagar,
Delhi – 110052
Ph.: 011-47034999, 9811692060
www.kalpazpublications.com
E-mail: kalpaz@hotmail.com

Distribution Network
gyanbooks.com
India, USA, Canada, UK, Australia, France

ISBN: 978-93-5722-279-2 (HB)

First Published, 2023

Printed at: Gyan Press, Delhi.

SUSTAINABLE TRANSITIONS IN INDIA
Editors: J.V. ARUN AND A. PREMKUMAR

Foreword

This volume, consisting of eighteen chapters, bearing on the topic of sustainable development goals, contributed mostly by young scholars, examines India's position in global sustainable development index and maps the direction of the progress in achieving the goals by 2030. It is an outcome of a national seminar, organized in the context of Covid-19 and makes an attempt to measure the setback to achieving SDGs by 2030 as a result of the prolonged pandemic and consequent disruption of educational, social, economic and other activities. Though this volume does not deal with all the seventeen goals, yet it touches upon them while remained focussed on poverty and hunger, health, education, agriculture, and water and sanitation. It also bears upon the issues of discrimination and equitable access to various resources, with a greater focus on caste than on gender.

While India remains at a low position, ranked 121 out of 163 countries, as per the global SDG Index 2022, its various efforts in achieving the SDG goals are appreciable. This book highlights that there are 11 SDG goals in which India remains poor performer and must give focussed attention to obtain them by 2030. It also underlines that Covid-19 crisis has increased the challenge, as it led to massive disruption in economic and other activities, leading to loss of livelihoods to the vast section of India's working population engaged in informal sectors. Many of them have got back to their normal economic activities, but a large number of them have suffered huge losses, difficult to recover, thereby affecting, India's progress in achieving various SDGs.

The important message of this book lies in emphasizing the urgency of giving more focussed attention to some vital goals, like SDG-2 (hunger), SDG-3 (good health and well-being), SDG-5 (gender equality), SDG-6 (water and sanitation), and SDG-8 (decent work and economic growth). Contrary to India's position that requires her to give priority to social sector development, the government has been shirking from the responsibility and inviting private sectors to take the lead role. The government budgetary allocation on social sector development remains stagnant in real terms and even declines in some sectors. As such there appears a gap between commitment to achieving SDGs and action on the ground, especially priority in resource allocation. A number of the essays in this book are based on empirical examination and provide a true picture of ground reality, based on micro-level study that is often obfuscated under the macro-picture, i.e., all India average

I appreciate the efforts of the editors and contributors in drawing our attention to India's position in achieving various SDGs by 2030 and underlining various challenges in this regard. I congratulate them for their endeavour in giving attention to such vital aspects, which often get lost amidst growth-fetishism and GDP-centric approach.

Prof. Ashok Pankaj
Council for Social Development
53, Lodi Estate
New Delhi

Preface

The United Nations Sustainable Development Goals (SDGs) are the universal agreement to end poverty in all its dimensions and craft an equal, just and secure world-for people, planet and prosperity by 2030. There are 17 Sustainable Development Goals and 169 measurable targets to be achieved by the nations. According to Global SDGs Index 2022, India secured 121 out of the 163 countries and the country is of the right track to attain SDGs. However, it continues to face major challenges to achieve 11 out of 17 goals which includes SDG-2 (zero hunger), SDG-3 (good health and well-being), SDG-5 (gender equality), SDG-6 (clean water and sanitation), SDG-8 (decent work and economic growth), SDG-9 (industry, innovation and infrastructure), SDG-11 (sustainable cities and communities), SDG-14 (life below water), SDG-15 (life on land), SDG-16 (peace, justice and strong institutions) and SDG-17 (partnerships for the goals). Not only this, the covid induced economic crisis is likely to impede India's progress in attaining all 17-Sustainable Development Goals (SDGs). During this crisis ridden period, mobilizing additional financial and technical support from international community to achieve SDGs will be a difficult proposition. In this critical juncture, India requires to identify innovative and affordable policies to ensure attainment of 2030 Agenda for sustainability and development. India before pandemic was not on track on most of the goals except SDG-1 (no poverty), SDG-6 (clean water & sanitation), SDG-8 (economic growth), and SDG-12 (climate action).

This book is a visionary document that would help in the revival of the economy and make India to achieve SDGs by 2030. The selected 18 papers in the book attempted to cover some of the UN SDGs. The selected papers were presented in the national seminar on "COVID-19 Crisis and Sustainable Development Goals in India: Pathways for Adaptation and Resilience" held on 19th and 20th October, 2022 organized by Department of Economics, Government Arts College for Men (Autonomous), Nandanam, Chennai and sponsored by Indian Council of Social Science Research, Southern Regional Centre (ICSSR-SRC), Hyderabad.

A total of 18 peer-reviewed chapters affords this book a strong thematic coverage. In the first chapter, Sha Hussain et al. attempted to address the problems of undernutrition among the children and budget allocation of the government to tackle malnutrition. Findings reveal that the pro-poor depend heavily on government allocations for child welfare schemes, while there was a massive disparity in the government's allocation to eradicate malnutrition during the fiscal year 2000-02 to 2019-21. The second chapter, from Nandhini et al. aim to find the determinant factors responsible for food wastage along the perishable food supply chain. The study highlights that technological and transportation factors, pandemic effects, and inefficient supply chains were responsible for food loss and waste in the perishable food supply chain. In third chapter, Kalaiarasi et al. examined the indicator of food price anomalies and food price inflation in India and also assessed the instability in price of major cereal and non-cereal crops during pandemic. The result revealed cereals and pulses showed low price instability whereas vegetables showed high price instability during 2019. Swathika in the chapter "Trends in Nanotechnology for sustainable food security" review the area of agricultural nanotechnology including nano fertilizers, nano pesticides and nano materials for seed quality improvement to address the challenges in food security and sustainability. In fifth chapter, Divya et al. examined the facilities and services of PHCs from the institutional perspective and also analyzed the perception of community in utilizing the health care services. The study revealed that, most of the rural PHCs are devoid of the basic infrastructural and other facilities compared to the urban PHCs and only few rural PHCs have maternal and referral facilities whereas no rural PHCs have inpatient facilities and this prevent the rural public from visiting the PHCs during their emergencies.

In the chapter, "A Sustainable Medical Practices in Treating Life Style Disorders", the authors, Zakiyyah Taj and Lakshmi evaluated the socio-economic factors influencing the choice of Traditional alternative medicine and conventional medicine. The study also analyzed the cost and effectiveness involved in both practices in treating lifestyle disorders. The chapter concludes that there was a significant difference seen in the age in the choice of medicine in treating their ailment compared to gender, income, qualification, and occupation, which shows insignificance.In seventh chapter, Sagar Kodi et al. critically analysed the impact of covid-19 on the health status of the Chenchus of Nallamala Hills based on primary data gathered from the field and aptly substantiated with secondary sources. The findings of the study reveal that, Chenchus find themselves in a very precarious position as they have been able to isolate themselves to prevent contracting the virus, yet, they were not left untouched by it. Even though very few positive cases and no deaths were recorded, they faced an uncertain future.

Deepa in the chapter "Effectiveness of Online Teaching and Learning Within the College during the Lockdown-A Study" explored the various modes of online teaching and learning during the pandemic period and also examined the teachers and learners' knowledge towards digital tools. The study found the transition from traditional to online pedagogy education has provided enough time for the nation particularly colleges and educational institutions. In ninth chapter, Shital and Ramesh examined how the expansion of higher education in India is moving towards the attainment of SDG 4 which focuses on inclusive, equitable, high-quality, and lifelong learning. The study found that, SDG-4 aiming for inclusive and equal access to higher education is successful in encouraging a greater number of people, irrespective of gender and socio-economic status, to continue their higher education after the completion of secondary level of education which enables them to not only improve their standard of living but also contributing towards the economic prosperity.

Arun and Premkumar in the chapter "Achieving Sustainable Sanitation Coverage in Urban Slum Settings" assessed the actual status of ODF. Programmes prior to 'Swachh Bharat Mission (SBM) – Clean IndiaMission', 2014 were futile and fell short of their target of achieving a clean India.The study concludes that effectiveness of post-ODF is measured in terms of attainment of safe and healthy society and equitable access to enhanced sanitation for each and every one. In chapter eleventh, Keerthika and Jayapriya examined the impact of COVID-19 on India's economy using the NSDP (Net State Domestic Product). The findings of the study revealed agriculture is the only sector which is having positive growth during and after COVID-19. A state-wise analysis shows how Indian states recovered from the lockdown and moved towards the Sustainable Development Goal of zero hunger, no poverty, and gender equality. In the chapter "Covid-19, Inequality, and Marginalised Sections in India", Karthick critically analysed the caste discrimination of marginalized sections in India during COVID-19. The author found that, over 50 percent of SC/STs faced difficulties in accessing non-covid medical facilities compared to general category and prevalence of disparity exist among social groups in accessing et online education.

In chapter thirteen, Smirthi et al. analyzed the trends on the productivity of rice in Tamil Nadu using the variables like temperature and rainfall from the period of 2005 – 2021. The study suggests to educate the common people about the devastating outcome of climate change and the urgent need to focus on sustainable lifestyle to have a future.In the chapter "Plastic Waste Economics and an alternative solution to this plastic menace", Megha et al. reviewed the statistics of plastic waste management for sustainable environment and the scope of bioplastics in the future. The study revealed that, use of plastic has increased twofold during the past 30 years as a result of the growth of new economies. Between 2000 and 2020, the production of plastics surged by a factor of two, reaching 460 million metric tonnes, or 3.4% of all worldwide greenhouse gas emissions.In fifteenth chapter, Jayapriya and Keerthika made an attempt to determine the impact of Soil Health Card (SHC) schemeon productivity and the constraints in implementation of the scheme. The study revealed that there was increased yield of 9.33 per cent while compared with non-beneficiaries with reduced fertilizer consumption. B:C ratio is also higher in SHC beneficiaries than non-beneficiaries.

In sixteenth chapter, Nijil Jacobi discusses many aspects of identification of destitute and planning proposals initiated by Kerala Institute of Local Administration (KILA) such as distinction between extreme poverty and poverty, and; interventions for the extreme poor by the state. The author highlights the importance of creating micro plans to improve the lives of the families that have been designated as the poorest in the state. In the chapter "Role of Caste in Access and Choice of Agricultural Credit", Suresh et al. identified the differences in access to credit of different social groups. Results indicate differences in loan disbursal among different social groups and different scales. The observed differences are more among the cooperative banks than the commercial banks. In the final chapter, Sangavi and Shobana examined the challenges faced by the vendors in dealing with digital payments in their day-to-day business transactions. The study indicated that most vendors were receptive to the idea of accepting payments through digital platforms, however the preference for having cash in hand still persists among the majority of the population.

We are grateful to the contributors for their insights and ICSSR-SRC for their financial support. We also like to thank the team at Gyan Books Pvt Ltd, New Delhi for their behind-the scenes help.

J. V. Arun
A. Premkumar

Contents

About the Editors

Dr. J.V. Arun, Assistant Professor of Economics at Government Arts College for Men (A), Nandanam, Chennai. He has fourteen years of teaching and research experiences. His research interest includes health and development economics. His PhD work is related to utilization of health care services and household health expenditure. He has good number of publications in reputed journals and books including Springer, Scopus-indexed, and UGC CARE listed. Further, he has edited a book titled *COVID-19 in Tamil Nadu: Repercussions and the Avenues Ahead*. He has carried out a research project, related to urban sanitation, funded by Tamil Nadu State Council for Higher Education (TANSCHE). Apart from that, he has an ongoing major research project on climate change funded by Indian Council of Social Science Research (ICSSR). Moreover, he has organized national level seminar on *COVID-19 crisis and SDGs in India* sponsored by ICSSR-SRC, Hyderabad. Also, he has been resourceful in various national and international economic forums by presenting research papers, and delivering invited lectures and special addresses. Articles written by him has appeared in leading Tamil newspapers and magazines.

Mr. A. Premkumar, PhD Research Scholar of Economics at Sacred Heart College (Autonomous), Tirupattur, Tamil Nadu. He has worked in various research projects at the Centre for Ecological Economics and Natural Resources (CEENR), Institute for Social and Economic Change (ISEC), Bangalore. The projects titled *Carbon Footprint and changing lifestyles of Indian Households* funded by Gesellschaft fur Internationale Zusammenarbeit (GIZ-International Cooperation) and *Climate Smart Livelihood and Socio-ecological Development of Biodiversity Hotspots of India* funded by Technology Information, Forecasting and Assessment Council (TIFAC), Government of India. He has written and published good number of articles in Scopus indexed, UGC Care listed journals and chapters in edited books published by Springer Nature in the area of Environment and Health. He also received doctoral fellowship from University Grants Commission (UGC), Government of India for his research.

List of Contributors

A. Premkumar, Research Scholar, Department of Economics, Sacred Heart College (Autonomous), Tirupattur, Tamil Nadu.

A. Rohini, Professor, Department of Agricultural and Rural Management, Centre for Agricultural Rural Development Studies, Tamil Nadu Agricultural University, Coimbatore.

Aswathy Ann Thomas, Department of Economics, Madras Christian College, Tambaram, Chennai.

B. Keerthika, Research Scholar, Department of Agricultural Economics, CARDS, TNAU, Coimbatore, India.

C. J. Lakshmi, Assistant Professor, Department of Economics, Mount Carmel College, Bengaluru.

D. Kalaiarasi, Research Scholar, Department of Agricultural and Rural Management, Centre for Agricultural Rural Development Studies, Tamil Nadu Agricultural University, Coimbatore.

D. Murugananthi, Assistant Professor, Department of Agricultural and Rural Management, Centre for Agricultural Rural Development Studies, Tamil Nadu Agricultural University, Coimbatore.

D. Sangavi, Department of Economics, Madras Christian College, Chennai.

Divya B Kunnath, Research Scholar, Department of Economics, Madras Christian College, Tambaram, Chennai.

J.V. Arun, Assistant Professor, Department of Economics, Government Arts College for Men (Autonomous), Nandanam, Chennai, Tamil Nadu.

K. Sadasivam, Registrar (i/c), Madurai Kamaraj University, Madurai, Tamil Nadu.

Kondaiah Nayak Malavathu, Independent Researcher, Washington D.C, USA.

M. Deepa, Assistant Professor, Department of History, Justice Basheer Ahmed Sayeed College for Women (Autonomous), Chennai.

M. Kamaraj, Associate Professor, Department of Biotechnology, Faculty of Science and Humanities, SRM Institute of Science and Technology (Ramapuram Campus), Chennai, Tamil Nadu.

M. Sagunthala, Research Scholar, Department of Environmental Economics, School of Economics, Madurai Kamaraj University, Madurai, Tamil Nadu.

Megha Mahajan, Research Scholar, Department of Biochemistry, College of Science and Humanities, SRM Institute of Science and Technology, Kattankulathur, Tamil Nadu.

Merlin Juliet Arul Thangam, Department of Economics, Madras Christian College, Tambaram, Chennai.

N. Swathika, Research Scholar, Centre for Agricultural Nano-technology, Tamil Nadu Agricultural University, Coimbatore.

Nijil Jacobi, Assistant Professor, Department of Economics, Sacred Heart College, Chalakudy, Kerala.

P. Ponsakthivel, Portfolio Manager, Yara Fertilizer India Pvt Ltd, Coimbatore, Tamil Nadu.

Prajitha Prakash, Research Scholar, Department of Environmental Economics, School of Economics, Madurai Kamaraj University, Madurai, Tamil Nadu.

PV. Bhargav, Research Scholar, Department of Humanities and Social Sciences, Indian Institute of Technology (IIT), Roorkee.

R. Jayapriya, Research Scholar, Department of Agricultural Economics, CARDS, TNAU, Coimbatore.

R. Shital, ICSSR Fellow, Research Scholar at Centre for Economic and Social Studies (CESS), Hyderabad.

Ramesh Annemalla, Research Scholar at Centre for Economic and Social Studies (CESS), Hyderabad.

S. D. Sivakumar, Professor, Department of Agricultural and Rural Management, Tamil Nadu Agricultural University, Coimbatore, Tamil Nadu.

S. Mobeen Farhana, Research Scholar, Department of Economics, Madras Christian College, Tambaram, Chennai.

S. Nandhini, Research Scholar, Department of Agricultural and Rural Management, Tamil Nadu Agricultural University, Coimbatore, Tamil Nadu.

S. Smirthi Rajya, Research Scholar, Department of Economics, Madras Christian College, Tambaram, Chennai.

Sagar Kodi, Research Scholar, Department of Anthropology, Pondicherry University, Puducherry.

Sha Hussain @ S.Yacob Khan, Research Scholar & ICSSR Fellow, Department of Environmental Economics, School of Economics, Madurai Kamaraj University, Madurai, Tamil Nadu.

Shobana Karthikeyan, Assistant Professor, Department of Economics, Madras Christian College, Chennai.

Suresh Govindapuram, Assistant Professor of Economics, IFMR GSB, KREA University, Andhra Pradesh.

T. Amose, Research Scholar, Department of Environmental Economics, School of Economics, Madurai Kamaraj University, Madurai, Tamil Nadu.

T.G. Nithya, Professor, Department of Biochemistry, College of Science and Humanities, SRM Institute of Science and Technology, Kattankulathur, Tamil Nadu.

Thottadi Bhavani Prasad, Research Scholar, Department of Humanities and Social Sciences, Indian Institute of Technology (IIT), Roorkee.

V. Jacquline Raja Bai, Associate Professor and Head, Department of Economics, Madras Christian College, Tambaram, Chennai.

V. Karthick, Research Manager, Tamil Nadu Household Panel Survey, Madras Institute of Development Studies (MIDS), Chennai.

Valerie Dkhar, Assistant Professor, Department of Anthropology, Pondicherry University, Puducherry.

Zakiyyah Taj, Department of Economics, Mount Carmel College, Bengaluru.

1

India's Action Plan on Reducing Malnutrition among Children

Sha Hussain @ S. Yacob Khan, K. Sadasivam, T. Amose, Prajitha Prakash and M. Sagunthala

Introduction

According to the World Food Programme, man-made conflicts, climate change and economic stagnation serves as a cause of malnutrition suffering for 135 million people at worldwide (UN-SDG, 2019). Correspondingly, the United Nations Inter-Agency Group for Child Mortality (UN IGME) estimates that child mortality and the global threat to Sustainable Development Goal (SDG) achievements are skewed by geographic and economic disparities. The UN IGME estimates that an average of 48 million children under the age of five may die before 2030, while an equal share will be new-borns. Of these deaths, 57 percent occur in Sub-Saharan Africa and 25 percent ensue in Southern parts of Asia. Moreover, eight million child deaths could be prevented between 2021 and 2030 by recognising the need to target the 54 less developed Sub-Saharan Africa and Southern Asia countries that have derailed from the SDG targets.Yet by 2020, five million babies would decease before reaching their fifth birthday (UN IGME, 2021).

According to Global Hunger Index report, India is ranked 101 out of 116 countries, indicating that the country has high levels of malnutrition and hunger-related issues. According to UNICEF report, in India,one in seventy-two children encounter stillbirth (loss of child life during delivery), one in fourty-seven decease during birth, and one in twenty-nine demises under age five. Besides, life loss of more than 50 percent of stillbirths, new-borns and children under five occur in the region of South-Asia (UNICEF global databases, 2021). Malnutrition is a major development concern and is closely linked to massive human and economic causes (Horton, 1999). Poor child health erodes social and economic gains, pushes countries into a vicious circle of poverty along with increased poor nutritional status, and high disease burden (Parker & Osmond, 1986).

The fourth round of the National Family Health Survey conducted in 2015-12 exposed that the prevalence of stunting among the children under-five was lowest in Kerala (13 percent) and highest in Uttar Pradesh (65 percent). Correspondingly, the incidence of underweight was lowest in Mizoram (7 percent) and highest in Jharkhand (67 percent) (NFHS-4 (2015–16)). Malnutrition is a country-wide phenomenon in India, and in terms of regional effects, malnutrition is concentrated in the least developed districts of the central, eastern and north-eastern regions, though the more developed districts of the western and southern regions. Singh et al. found that the socio-economic inequality still exists as undernutrition in India (Singh S et al. 2019).

However, most of the studies focus on where India stands in terms of malnutrition, district-wise malnutrition and child healthcare. There are no studies that integrate the status of malnourished children and the country's progress towards reducing the impact of malnutrition. Due to the lack of reflection on the topic, *this study aims to address the prevalence of malnutrition among under-five children in India and the government's budget allocation with an effort to tackle the problem of child malnutrition. In particular, the study focuses on addressing nutritional needs, one of the goals of Sustainable Development Goal – Zero Hunger.*

Methods and Materials

The secondary data for the study on undernutrition among the children under five years of age were retrieved from the Global Hunger Index 2021 report and the Food and Agriculture Organization (FAO) dataset with regular intervals that covering the period from 2000-02 to 2019-21. India's mission to end hunger and budget allocation for reducing malnutrition has been collected from the reports of Union Budget of Ministry of Finance: 2019-20 & 2020-21, Demand for Grants 2020-21 and Press Information Bureau (PIB). Moreover, data on malnutrition rate, child wasting, child stunting and under-five mortality rate are available for 2000-02, 2004-06, 2010-12 and 2019-21. Hence, the study considered the specified time period and used for analysis. Percentage change calculation method is used to determine the inter-decadal difference and change in child malnutrition.

Results and Discussion

In the field of hunger in 2000-02, India maintained a very alarming score (38.8%). Meanwhile, through massive implementation of nutrition programmes (ex. ICDS, MDM, NFSM), India has weathered the situation and reduced malnutrition from 37.4 percent in 2004-06 to 28.8 percent in 2010-12. On the contrary, it is clear from Table 1 that the India's meagre budget allocation for a decade (2010 to 2020) has reduced malnutrition by just 4.62 percent between 2010-12 and 2019-21. However, India is still in a serious ranking 101 out of 116 countries (GHI, 2021). Besides, the level of undernutrition increased from 18.4 percent in 2000-02 to 19.6 percent in 2004-06, while the allocation has changed to 15 percent and 15.3 percent in subsequent years. Although there is a slight change from 2010-12 to 2019-21, it is still positive (1.98%). In comparison, stunting is a more serious problem among the children under five than wasting. From a peak of 54.2 percent in 2000-02, it has declined to 34.7 percent in 2019-21, indicating that India is still struggling to combat stunting among children under-five. Also, under-five mortality rate shows a positive change of 41.86 percent between 5.2 percent in 2010-12 and 3.4 percent in 2019-21. Saha

S and Singh R (2021) stated that India's child malnutrition crisis is largely due to historical antecedents of poverty, inequality and food insecurity (Saha S and Singh R, 2021).

Table 1: Nutritional deficiency in children under five age group (in %)

Description	2000-02	2004-06	2010-12	2019-21	% Δ 2010-12 & 2019-21
GHI Country Level Scores	38.8	37.4	28.8	27.5	4.62
Proportion of undernourished	18.4	19.6	15.0	15.3	1.98
Wasting in children (<5)	17.1	20.0	15.1	17.3	13.58
Stunting in children (<5)	54.2	47.8	38.7	34.7	10.90
Under-five mortality rate	9.2	7.1	5.2	3.4	41.86

Source: Global Hunger Index Report, 2021

Table 2 express the proportion of malnutrition among the children under five in millions. According to Food and Agriculture Organization (FAO) data, the number of undernourished, stunted and overweight people in India during 2000-02 was found to be 198.3, 63.7 and 4.2 million. In subsequent years, malnutrition among under-five children increased by 24.96 percent in 2004-06, whereas stunting and overweight decreased by 2.2 percent and 8.3 percent, correspondingly. In 2010-12, 193.1, 53.6 and 3.1 million children were affected by malnutrition, stunting and overweight. Between 2010-12 and 2019-21, malnutrition among under-five children in India has increased by 15 percent, followed by stunting and overweight by 39.02 percent and 34 percent, respectively. The fourth round of the National Family Health Survey (NFHS-4) conducted in 2015-16 found that malnutrition rates are high, especially in rural areas. However, critics point to low budget allocation and underutilization as the main reasons for the rise in undernourishment among under-five children (Chadha S, 2022; Roy E, 2022; Jayashree B & Gopinath R, 2020).

Table 2: Malnutrition Defect of Indian Populace (in million)

Item	2000-02	2004-06	2010-12	2019-21	% Δ 2010-12 & 2019-21
No. of people undernourished (3-year average)	198.3	247.8	193.1	224.3	14.95
No. of children < 5 years of age who are stunted	63.7	62.3	53.6	36.1	39.02
No. of children < 5 years of age who are overweight	4.2	3.85	3.1	2.2	33.96

Source: Food and Agriculture Organization of the United Nations (FAOSTAT)

Malnutrition among children and pregnant women accounts for 15 percent of all diseases in India (NFHS-4, 2015-16). As a priority, India has implemented various major programmes to eliminate the prevalence of malnutrition, whereas the implemented schemes date back to the 1970s to present (Table 3). In 1975, India introduced the ICDS to improve the nutritional status of children and to educate feeding women about the nutritional needs of children. In FY 2019-20, the Government of India allocated Rs 17,705 crore, which was reduced by 2.6 percent to Rs 17,252 crore in FY 2020-21. In 2007, the NFSM was implemented to address food and nutrition security and ensure accessibility, adequacy and affordability of food and food prices. In FY 2019-20, India has come up with a budget of Rs 1,500 crore and the scheme has witnessed a growth of 32.45 percent with Rs 2,000 crore in FY 2020-21. In contrast, the flagship programme POSHAN Abhiyan (National Nutrition Mission), which was launched in 2018 and focuses on children, adolescent girls, pregnant women and lactating women, has seen a decrease of 66.6 percent from the previous fiscal year 2019-20 and 83.8 percent of the revised budget as compared to the budget estimate. Lastly, the MDM programme later renamed PM-POSHAN, is a major source of food for school-going children. From FY 2019-20 to FY 2021-22, the budget increased by 32.8 percent from Rs 9,690 crore to Rs 12,870 crore. Later, the budget estimates of the MDM programme

Table 3: India's Missions for Ending Hunger as a Priority

Name of the Mission	Year of Implementation	Mission's Goal and Target groups	Budget Allocation (in crore)	
			2019-20	2020-21
Mid-Day Meal (MDM) Scheme (PM POSHAN)	Launched in 1995 and altered as PM-POSHAN (Prime Minister's Overarching Scheme for Holistic Nutrition) in 2021	to provide food to all children enrolled in government aided schools, local government schools and special training centres.	9,690**	12,870**
National Nutrition Mission (POSHAN Abhiyaan)	2018	to reduce malnutrition and improve the nutritional status of children 0-6 years, adolescent girls, pregnant women and lactating women.	1,800**	3,700* - 600**
Zero Hunger Programme	2017	to double agricultural production and incomes of small-scale farmers, eliminate food waste, ensure sustainable food production and food availability worldwide, and reduce stunting among children aged less than 2.	--	--
National Food Security Mission (NFSM)	2007	to provide food and nutrition security, ensuring that people have access to adequate and quality food at affordable prices.	1,510**	2,000**
Integrated Child Development Services (ICDS)	1975	to improve the nutritional status of the children, reduce infant mortality, and advance the mothers capability to satisfy the children's nutritional need.	17,705**	17,252**

Source: Union Budget of Ministry of Finance: 2019-20 & 2020-21, Demand for Grants 2020-21& Press Information Bureau (PIB).

Note: * - Budget Estimate (BE) & ** - Revised Estimate (RE)

for the financial year 2021-22 have been reduced from 11.5 to 10.23 thousand crores respectively (Union Budget, 2021-22).

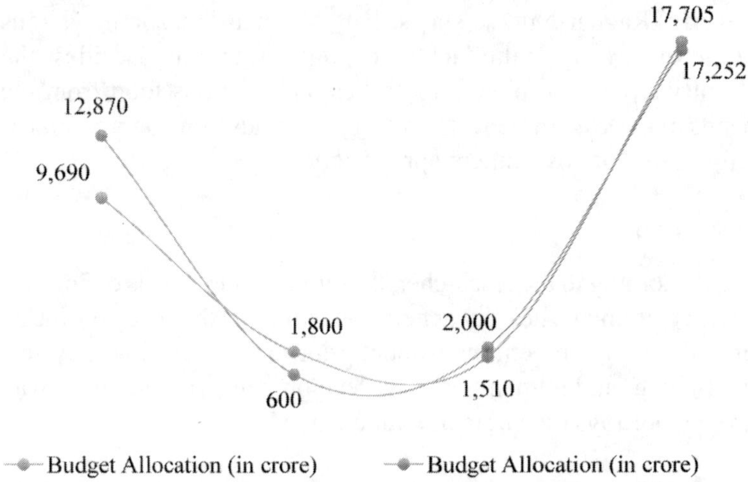

Fig. 1: India's Budget Allocation to Reduce Child Malnutrition.

Theoretical Perceptions

In general, theory graph focus on the theme and thoughts of the authors on an innovative idea needed for a global trend. But here the study refers to such economic theoretical graph (Vicious Circle of Poverty) that bring innovative picture to identify the centers of study theme to influence recommendations for policy-makers related to malnutrition (Amose T and Jeyakumar K R, 2018). Thereby, Figure 2 demonstrates how and why the cycle of malnutrition begins and ends.

Poverty is the root cause for food insecurity and nutrition deficiency (Müller O & Krawinkel M, 2005). Moreover, there is a two-way link between malnutrition and poverty, each fuelling the other, creating a vicious cycle (Siddiqui F et al. 2020). Thereby, this study adopts the concept of the vicious circle of poverty and adjusts the variable accordingly.

Meaning

The poverty cycle proposes that 'poverty is the root of poverty'.Ragnar Nurkse states that, "the main reason of vicious circle of poverty is the lack of capital formation." Besides, the country's pro-poor population relies on nutritious food from the hands of the government. Therefore, this study considers capital as a formation of government appropriation.

Definition

According to the researcher, the circle of nutrition is defined as, "The creation of a healthy generation will sows the seeds for future growth; if government investment prioritises – nutrition, physical well-being, and human resource development, and the other way around portrays the circle of malnutrition."

Assumptions

The vicious circle of malnutrition is based on the following assumptions.

The basic nutrition of the grassroots is determined by government allocations for nutrition related schemes.

The future of domestic production today depends on the nutrition that fills the bellies of the nation's children.

For the past ten years, the nation's sectorial financial allocation had neglected people's health concerns.

Theoretical explanation

Resources are underdeveloped and people in less developed countries are economically disadvantaged (Ehrlich O, 1958). According to this statement, Fig. 2 confirms that the occurrence of imperfect or scarce resource supply prevents optimal utilization of resources and consequently underdevelopment and this leads to child malnutrition. Nevertheless, the theoretical framework of undernutrition suggests that a small spread of the budget induces an underutilization of resources. In addition, low consumption increases inequality and food insecurity among children under five, leading to malnutrition, stunting and wasting. Malnutrition among children creates an unhealthy workforce in the future, leading to lower productivity and lower gross output.

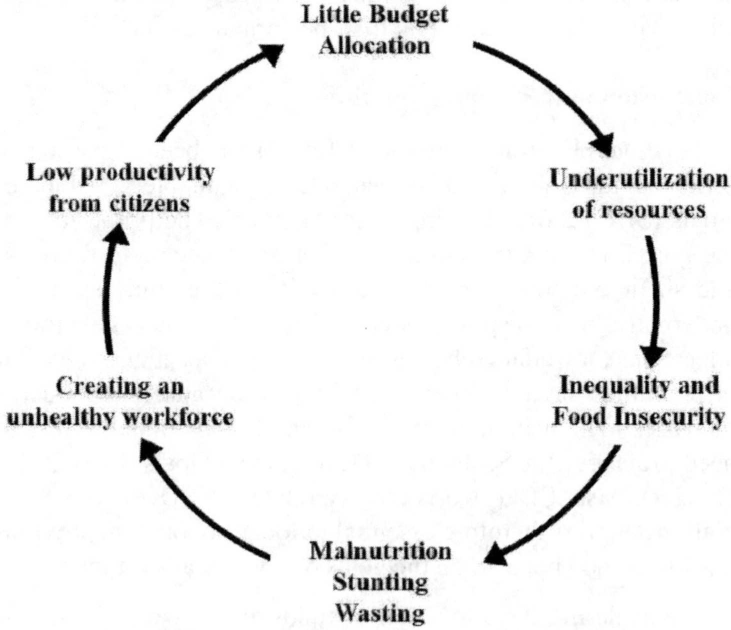

Fig. 2: The Vicious Circle: Malnutrition side.

As the number of malnourished children increases, on the other hand, the budget keeps decreasing. It is clear from Table 3 and Fig.1 that the allocation for the ICDS programme for the financial year 2019-21 has been reduced by 2.6 percent respectively. Moreover, India faces a decline of 4 percent in GDP and 8 percent in productivity every year (GHI, 2021). Henceforth, the need for India to be '*on course*' to prevent this number from increasing.

Furthermore, malnutrition adversely affects the physical and mental capacity of individuals; This hinders productivity, impoverishes them and their respective countries (Siddique F,

2020). So far, deficiencies in quotas can be rectified in two ways. First, apart from diverting allocation, dispersal of funds into productive channels can be of great help in breaking the vicious cycle of malnutrition. Second, short-term and long-term investment policies should be coordinated. With a short-term investment, people can get the things they need at a reasonable price, which will have a positive impact on their performance.

Conclusion and Recommendations

The novel corona virus (COVID-19) has been a disaster for achieving the United Nations-mandated Sustainable Development Goals (SDG) 2 of achieving 'Zero Hunger'. The prime focus of the goal 2 is – ending hunger and ensuring access to nutritious and sufficient food for all. The results of the study show that undernutrition of the pro-poor is heavily dependent on government allocations for welfare schemes. In particular, allocations for welfare programmes have declined over the past decade.This reduction increases malnutrition among children, thereby undermining the need to achieve the Sustainable Development Goals (Goal 2: Zero Hunger). Also, SDG two is closely related to SDG one: poverty eradication. It is therefore essential to focus not only on providing nutritious food but also on the goals of poverty alleviation.

Considering the allocation of funds and impact on children, the study recommends collaboration and joint action is needed from Ministries such as Rural Development, Public Distribution and Civil Supply, Health and Family Welfare, Women and Child Development, Drinking Water and Sanitation, Agriculture, Tribal Affairs. Minority Affairs. Because it is essential to eradicate poverty and malnutrition among children. On the other hand, there is a need to reduce the gaps in data, invest more in social welfare programmes, monitor the budget allocated to the specific programme and track the flow of funds in rural areas to break the vicious cycle of malnutrition.

Acknowledgements

The corresponding author of the paper, Sha Hussain @ Yacob Khan is an ICSSR Doctoral Fellow. Hence, the author acknowledge Indian Council of Social Science Research (ICSSR) Doctoral Fellowship for the financial support to carrying out the research. This article is largely an outcome of the author's doctoral work sponsored by ICSSR. However, the responsibility for the facts stated, opinions expressed and the conclusions drawn is entirely that of the authors.

References

Amose, T and Jeyakumar, K R. (2018). Contribution of Energy to Food Security and Economic Analysis to Supply, *Scholars Academic and Scientific Publishers* (SAS Publishers), 6(1C): 125-134. DOI: 10.21276/sjahss.2018.6.1.22

Anitha, S and Botha, R and Kane-Potaka, J and Givens, I D and Rajendran, A and Tsusaka, T W and Bhandari, R K (2021) Can Millet Consumption Help Manage Hyperlipidemia and Obesity?: A Systematic Review and Meta-Analysis. *Frontiers in Nutrition* (TSI), 8 (700778). ISSN 2296-861X. Retrieved from: http://oar.icrisat.org/11995/1/21.%20Lipid%20 manucript_2021.pdf

Barker, DJ, Osmond, C. (1986). Infant mortality, childhood nutrition, and ischaemic heart disease in England and Wales. *Lancet*. 1(8489):1077-81. doi: https://doi.org/10.1016/s0140-6736(86)91340-1

Chadha, S (2022). Include eggs fruits in Mid-Day-Meals strengthen Poshan 2.0: Budget 2022 must increase allocation towards child nutrition, *The Times of India*. Retrieved from: https://timesofindia.indiatimes.com/business/india-business/include-eggs-fruits-in-mid-day-meals-strengthen-poshan-2-0-budget-2022-must-increase-allocation-towards-child-nutrition/articleshow/89110982.cms

Ehrlich, O. (1958). Economic Development: Theory, History, Policy. By Gerald, M. Meier and Robert, E. Baldwin. New York: John Wiley & Sons, Inc., 1957. pp. xx, 588. $8.50. *The Journal of Economic History*, 18(1), 74-74. DOI: 10.1017/S0022050700088860

Food and Agriculture Organization – FoodSecurity Indicators. Retrieved from: https://fenixservices.fao.org/faostat/static/documents/FS/Descriptions_and_Metadata.xlsx

Global Nutrition Report.(2021). Retrieved from: https://globalnutritionreport.org/resources/nutrition-profiles/asia/southern-asia/india/

Horton, Susan. (1999). Opportunities for Investments in Nutrition in Low-income Asia. *Asian Development Bank*. http://hdl.handle.net/11540/5393.

International Institute for Population Sciences (IIPS) and ICF. 2017. National Family Health Survey (NFHS-4), 2015-16: India. Mumbai: *IIPS*. Retrieved from: http://rchiips.org/nfhs/NFHS-4Reports/India.pdf

Jayashree, B and Gopinath, R. (2020). Nutrition and the Budget's fine print, *The Hindu*. Retrieved from: https://www.thehindu.com/opinion/lead/nutrition-and-the-budgets-fine-print/article30803731.ece?homepage=true

Kumar, A, Kumari, D and Singh, A. (2015). Increasing socioeconomic inequality in childhood undernutrition in urban India: trends between 1992–93, 1998–99 and 2005–06, *Health Policy and Planning*, 30(8),1003–1016, https://doi.org/10.1093/heapol/czu104

Müller, O andKrawinkel, M. (2005). Malnutrition and health in developing countries, *Canadian Medical Association Journal*, 173 (3): 279-286; DOI: 10.1503/cmaj.050342

Roy, E (2022). Budget for WCD hiked, but no increase in funds for its flagship scheme, *The Indian Express*. Retrieved from: https://indianexpress.com/article/india/budget-for-wcd-hiked-but-no-increase-in-funds-for-its-flagship-scheme-7752297/#:~:text=%E2%80%9CFor%20instance%2C%20in%20FY%202019,21%20to%20Rs%2-017%2C252%20crore

Saha, S and Singh, R. (2021). Child malnutrition in India: A systemic failure, *Down-to-earth*. Retrieved from: https://www.downtoearth.org.in/blog/health/child-malnutrition-in-india-a-systemic-failure-76507#:~:text=The%20crisis%20of%20child%20malnutrition,income%20have%20fared%20much%20better.

Setboonsarng, S (2005). Child malnutrition as a poverty indicator: An evaluation in the context of different development interventions in Indonesia, ADBI Discussion Paper No. 21, *Asian Development Bank Institute* (ADBI).Retrieved from: http://hdl.handle.net/10419/53441

Siddiqui, F, Salam, A R, Lassi, S Z and Das, K J. (2020). The Intertwined Relationship Between Malnutrition and Poverty, *Frontiers in Public Health*. DOI: 10.3389/fpubh.2020.00453

Singh, S, Srivastava, S and Upadhyay, K A. (2019). Socio-economic inequality in malnutrition among children in India: an analysis of 640 districts from National Family Health Survey (2015–16), *International Journal for Equity in Health*, 18(203), 1-9. DOI: https://equityhealthj.biomedcentral.com/articles/10.1186/s12939-019-1093-0

UN Inter-agency Group for Child Mortality Estimation. (2021). Levels & Trends in Child Mortality: Report 2021 – *Estimates developed by the UN Inter-agency Group for Child Mortality Estimation*. URL: https://data.unicef.org/resources/levels-and-trends-in-child-mortality/

United Nations Sustainable Development Goal. (2019). *United Nations: Goal 2: Zero Hunger*. URL: https://www.un.org/sustainabledevelopment/hunger/

2

Towards Zero Wastage of Perishables – Food Loss and Waste in the Perishable Food Supply Chain

S. Nandhini, S.D. Sivakumar and *P. Ponsakthivel*

Introduction

India is the second largest producer of fruits and vegetables, next to China. India produced 88.97 million metric tonnes of fruits and 162.98 million metric tonnes of vegetables in 2014, constituting around 12.6 percent and 14 percent of the total world production of fruits and vegetables, respectively. Grapes, oranges, pomegranates, bananas, and mangoes account for a larger portion of fruits exported from the country. At the same time, onions, mixed vegetables, potatoes, tomatoes, and green chili contribute primarily to the vegetable export basket. Therefore, perishable commodities like fruits and vegetables contribute a major share in production and export from the country. Studies reported that around 1.3 billion tonnes of food are wasted and loosed globally due to the ineffective supply chain structure and consumer behaviour on food wastage Aramyan et al. (2020). Food loss and food wastage are two different concepts. Irani et al. (2018) defined food loss as the loss that occurs from primary supply chain stages of food, from planting to processing. Contrarily, food wastage occurs in the later stages and is typically brought on by consumer behaviour. Any food that is produced but not eaten by people is referred to as food waste, (Bilska et al. 2016). Typically, it is believed that food loss or waste refers to the amount of food that is thrown away or goes uneaten.

Food waste happens at every stage of the food supply chain in both developed and underdeveloped nations, but the reasons for their wastage are different. Around 18 percent of fruits and vegetables produced is wasted from post-harvest processing until it reaches the consumer. (FASAR, 2014). In developing countries like India, there is a significant amount of waste along the supply chain of perishable fresh food, fruits, and vegetables, right from production to harvest until the point of consumption. Many types of food loss can occur along the supply chain, such as spills, threshing degradation, and mechanical damage. Food quality, safety, food security and economic development are all directly impacted by food losses and directly and negatively affect both farmers and consumers (Gustavsson et al. 2011).

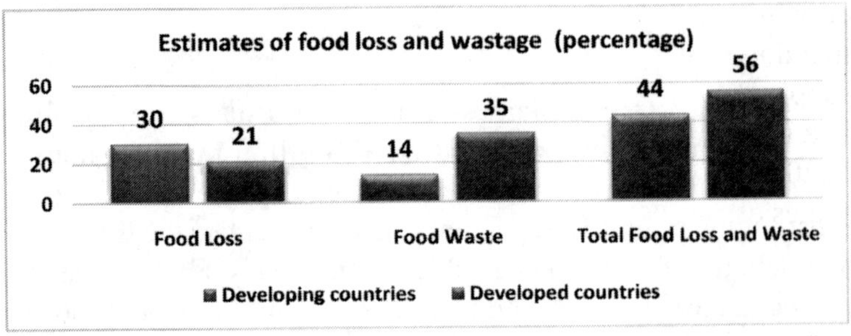

Fig 1: Food Loss and wastage in developing and developed countries.

Source: Ishangulyyev et al. (2019).

Food wastage is speculated to be intimately related to ineffective and inefficient food supply chain and consumption management. Businesses can assure operational efficiency through sustainable supply chain management (SSCM), which also increases their capacity to succeed in logistical and resource-intensive operations. It also aids businesses in achieving the United Nations Sustainable Development Goals (SDGs), particularly the "triple-bottom-line concept's" three dimensions of sustainability in terms of the economy, society, and the environment. Around 12 percent of the world's population is undernourished. In this situation, it is essential to comprehend the agricultural food supply chain issues and concentrate on boosting productivity. Figure 1 explains about the food Loss and wastage occurs in developing and developed countries.

Perishable food products are sold without proper packaging and primary processing due to a lack of necessary infrastructure, equipment, and packaging techniques. From a resilience viewpoint, (Costa et al. 2022) did research on food waste in Brazil. They concluded that especially in the Covid-19 circumstance, leadership, communication, knowledge management, innovation, and supply chain design greatly reduced food loss. However, a shortage of funds and layoffs among stakeholders may result in more food wastage. Covid-19 has increased food waste due to delays in harvesting and transportation. Food is regularly wasted in numerous circumstances because it cannot reach the ultimate users due to lockdown concerns. Covid-19 has caused food wastage in the supply chain where millions of gallons of milk are wasted, resulting in an economic slump and increased food scarcity (Corkery et al. 2020). Figure 2 explains the food loss and wastage in different stages of perishable supply chain.

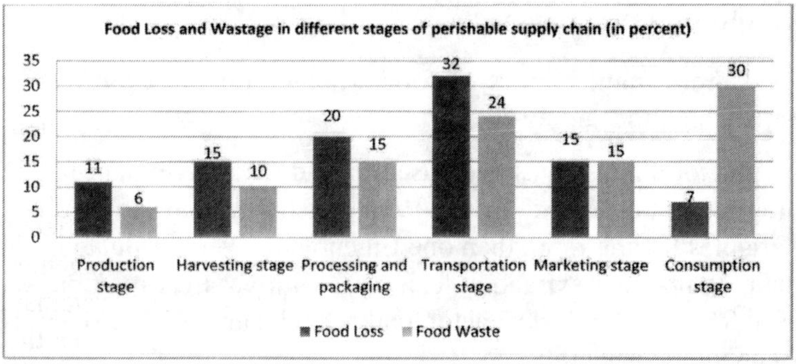

Fig 2: Food loss and wastage in different stages of perishable supply chain.

Source: Calculated primary data.

Research methodology

The study area selected for this research is Coimbatore district of Tamil Nadu, India in order to study the determinant factors of food loss and wastage in the perishable supply chain. Balaji and Arshinder (2016) depicted the causes of food waste in the perishable food supply chain. The major stakeholders involved in the perishable food supply chain includes fruits and vegetable farmers, traders, distributors and retail business. A total of 60 interviews were conducted, and qualitative insights were gathered from all the stakeholders with semi-structured interview schedule during May, 2022. The interviewees were 15 vegetable farmers (mostly engaged in growing tomatoes, cauliflower,and brinjal), 15 Fruit growers (mangoes, Guava, Jamun plum, and bananas), 10 traders, 10 distributors, and 10 retailers. The interviews concentrated on earlier food loss and food wastage that occurred in the perishable supply chain. This research work entirely focused on the following objectives:

- To identify the determinant factors of food wastage during Covid-19 in perishable food supply chain
- To suggest corrective measures to minimize the perishable food wastage and food loss in the perishable food supply chain

Results and Discussion

Determinant factors of food wastage during Covid-19 in perishable food supply chain.

Factor analysis has been used to find the determinant factors of food wastage during Covid-19 in perishable food supply chain. Variables having more than one Eigen value were evaluated. The main component extraction technique with varimax rotation was used for factor analysis. Factor loadings with greater than 0.5 were taken into consideration.

KMO and Bartlett's Test

Table 1: KMO and Bartlett's Test

Kaiser-Meyer-Olkin Sampling Adequacy Measure		0.623
Bartlett's Test of Sphericity	Approx. Chi-Square	36.215
	Df	**45**
	Sig.	**.000**

The above Table 1 shows that the KMO test value is 0.623, which is more than 0.5, indicating that the sample is adequate and suitable for conducting the factor analysis. In Bartlett's test, the chi-square statistic is approximately 36.215 with 45 degrees of freedom, which is significant at 0.000 levels.

Table 2 shows that four components have an Eigenvalue of 1.685, 1.528, 1.257, and 1.086 having a variance of 16.849, 15.275, 12.575, and 10.858, respectively. These four factors explain about 55.5 percent of the variance. In the rotated component matrix, ten variables taken in this study were grouped under four factors with factor loadings greater than or equal to 0.4. The identified variables with each factor were named and represented in Tables 3 and 4.

The study indicates that a lack of technological factors like traceability systems, processing techniques, and further lack of infrastructure facilities may increase food wastage in the perishable food supply chain. The most important factor for food wastage may be due to the pandemic as the perishable food products were overstocked in retail shops due to the lockdown effect. In addition to that, insufficient logistics infrastructure and non-availability of refrigerated trucks, lack of cold storage facilities may lead the food supply chain to have a huge quantity of perishable food wastage. Lack of coordination among supply chain actors may cause food wastage as there occurs a distortion of information regarding demand and supply for products and improper inventory management practices. The supply chain handling perishable food products should not have more intermediaries to avoid food wastage. Improper package practices in perishables may cause

Table 2: Total Variance Explained

Component	Initial Eigenvalue			Extraction Sums of Squared Loadings			Rotation Sums of Squared Loadings		
	Total	% of Variance	Cumulative %	Total	% of Variance	Cumulative %	Total	% of Variance	Cumulative %
1	1.685	16.849	16.849	1.685	16.849	16.849	1.575	15.751	15.751
2	1.528	15.275	32.124	1.528	15.275	32.124	1.485	14.848	30.599
3	1.257	12.575	44.699	1.257	12.575	44.699	1.338	13.378	43.977
4	1.086	10.856	55.555	1.086	10.856	55.555	1.158	11.578	55.555
5	.946	9.459	65.014						
6	.853	8.526	73.540						
7	.771	7.714	81.254						
8	.734	7.339	88.594						
9	.627	6.272	94.866						
10	.513	5.134	100.000						

Extraction Method: PCA

Table 3: Rotated Component Matrix

Variables	**Components**			
	1	**2**	**3**	**4**
Lack of contemporary processing techniques	.731			
Absence of traceability system	.701			
Inappropriate demand forecasting	.598			
Dearth of refrigerated trucks		.514		
Insufficient logistics infrastructure		.607		
inadequate cold storage facilities		.737		
Lack of coordination among supply chain actors			.899	
Involves large number of intermediaries in supply chain			.444	
Improper package of perishable products			.664	
Contamination of perishables due to prolonged storage in retail shop				.758
Extraction Method: PCA. **Rotation Method**: Varimax with Kaiser Normalization.				
a. Rotation converged in 6 iterations.				

Table 4: Factors determining the wastages of perishable in food supply chain

Factors	Items	Factor loadings
Technological factors	Absence of traceability system	.701
	Lack of contemporary processing techniques	.731
	Inappropriate demand forecasting techniques	.598
Pandemic factors	Contamination of perishables due to prolonged storage in the retail shop	.758
Transportation factors	Insufficient logistics infrastructure	.607
	A dearth of refrigerated trucks	.514
	inadequate cold storage facilities	.737
Inefficient supply chain factors	Lack of coordination among supply chain actors	.899
	Involves a large number of intermediaries in the supply chain	.444
	Improper packaging of perishable products	.664

damage to the products that might lead to loss and wastage in the food supply chain.

Conclusion

Food wastage is one of the most important issues happening in the world especially in perishable commodities. Each forkful of food wasted is responsible for greenhouse gas emissions. One-third of greenhouse gas emissions come from food loss and wastage specifically from perishable commodities. The findings reveal that food wastage has a major impact on climate change compared to food losses. The carbon footprint of food wastage occurs more in the consumption phase. Therefore, consumer awareness should be created among consumers regarding food loss and waste that leads to economic and environmental issues that directly impact society. Inadequate infrastructure facilities like insufficient cold storage facilities and improper packaging of goods were found to impact transportation and lead to more wastage of perishables negatively. Additionally, recycled packaging materials can be utilized to save money which reduces the risk of product damage. More focus on improving infrastructure facilities in the food supply chain by government and non-government organizations can help in reducing food loss and waste. Creating more awareness among entrepreneurs about the Indian government's initiatives like a five-year tax exemption for new SMEs would open up the potential to locate export markets for perishables. To ensure food safety and quality, food traceability is now a requirement, allowing for the tracking and tracing of food along the entire food supply chain.

Further data availability regarding food wastage at global and national levels is very low; hence, country and sector-wise food wastage details need to be made available for researchers for further studies. In addition, researchers need to give more consistent attention to undertake a research on the carbon footprint of perishables products associated with food wastage. Studies on consumer interest in purchasing products with a lower carbon footprint to ensure sustainable consumption can also be recommended.

References

Aramyan, L., Grainger, M., Logatcheva, K., Piras, S., Setti, M., Stewart, G., &Vittuari, M. (2020). Food waste reduction in supply chains through innovations: a review. *Measuring Business Excellence*.

Balaji, M., &Arshinder, K. (2016). Modeling the causes of food wastage in Indian perishable food supply chain. *Resources, Conservation and Recycling, 114*, 153-167.

Bilska, B., Wrzosek, M., Kołożyn-Krajewska, D., & Krajewski, K. (2016). Risk of food losses and potential of food recovery for social purposes. *Waste Management, 52*, 269-277.

Corkery, M., Yaffe-Bellany, D., & NY, T. (2020). Dumped Milk, Smashed Eggs, Plowed Vegetables: Food Waste of the Pandemic. U. *The New York Times*.

de Oliveira Costa, F. H., de Moraes, C. C., da Silva, A. L., Delai, I., Chaudhuri, A., & Pereira, C. R. (2022). Does resilience reduce food waste? Analysis of Brazilian supplier-retailer dyad. *Journal of Cleaner Production, 338*, 130488.

FASAR – Food and Agribusiness Strategic Advisory & Research YES bank. (2014).Fruits & Vegetables Availability Maps of India, Retrieved from the website of Ministry of Food Processing Industries, Government of India. www.mofpi.nic.in. (accessed 09.07.22).

Gustafsson, J., Cederberg, C., Sonesson, U., & Emanuelsson, A. (2013). The methodology of the FAO study: Global Food Losses and Food Waste-extent, causes and prevention" FAO, 2011.

Irani, Z., Sharif, A. M., Lee, H., Aktas, E., Topaloğlu, Z., van't Wout, T., & Huda, S. (2018). Managing food security through food waste and loss: Small data to big data. *Computers & Operations Research, 98*, 367-383.

Ishangulyyev, R., Kim, S., & Lee, S. H. (2019). Understanding food loss and waste—why are we losing and wasting food?. *Foods*, *8*(8), 297.

Sethi, S. Prakash, and Donald H. Schepers. "United Nations global compact: The promise–performance gap." *Journal of Business Ethics* 122, no. 2 (2014): 193-208.

3

Price Instability in Indian Commodity Market during Covid-19

D. Kalaiarasi, A. Rohini and *D. Murugananthi*

Introduction

The novel coronavirus outbreak has dispensed the Indian economy a fatal blow. Health professionals think that till a vaccine is developed and a mass vaccination programme is implemented, social exclusion, isolation, and restricting or banning travel across national and international boundaries are required controls to prevent additional outbreaks.This nationwide lockdown is the most restrictive measure ever taken by any nation to stop the spread of the disease, in addition to being unprecedented (Times of India, 2020).

In northern and central India, the rabi crop harvest—which includes wheat, pulses, and oilseeds—occurred at the same time as the shutdown. Because of a strong monsoon, most crops produced records during the harvest season (Moneycontrol, 2020). Production abundance often indicates price stability. But the epidemic has an impact on pricing in two ways: by disrupting agricultural supply networks and altering consumer demand. Despite the lifting of the restriction on the interstate movement of necessities, there remaineda interruption in supply chains of agricultural goods.

Since lockdown, markets have either been shuttered or operating at reduced capacity, which has hindered the flow of fresh product from farmers to customers. Commodity transportation across State lines was limited. The daily truck traffic was only 10 percent of pre-lockout levels for a few weeks following the initial lockdown. According to reports, agricultural supply stores remained closed. Timely seeding of summer crop and forthcoming kharif crop were both jeopardised by all of these reasons taken together (Kaur, 2020). Supply chain was severely constrained for many days following first shutdown despite clearance to control Agricultural Produce Market Committee (APMC) mandis, vehicles, fertilizer stores. Even several farmworkers claimed to be ignorant of the exemption for farm employment. The widespread migration of migrant labourers to their home towns, which resulted in a manpower shortage on farms, exacerbated this supply chain issue (Ellis-Petersen and Chaurasia, 2020). In the aftermath of this migration, there were also looming worries of food and starvation (Adhikari et al. 2020).

On demand side, due to panic huge consumers bought food items with long shelf life because they thought the lockdown would last for longer period. As a result, there was a spike in demand for commodities including pulses, onions, and potatoes, which led to a sharp increase in their prices. Because the large consumers groups viz., hotels, restaurants, hostel sand dhabas were unable to purchase in quantity, the demand for perishable fruits and vegetables substantially decreased.

The government's attempts to control the worsening situation were focused on reducing the supply chain interruptions and expanding access to food. The government released organizations involved in the purchase of agricultural products from the lockdown, including the minimum support price operations, mandis operation, farmer-run farms, fertilizer, pesticide, and seed manufacturing and packaging facilities, intrastate and interstate movement of harvesting and sowing-related machinery (Tripathi, 2020). On March 26, 2020, FM, Nirmala Sitharaman unveiled a socioeconomic package worth US$ 22.6bn to food and income security (Beniwal and Srivastava, 2020). This included actions for more wheat and rice to be provided to households enrolled in the Public Distribution System (PDS) each month over the following three months, as well as steps for direct cash transfers under popular government programmes. Later, this allocation was expanded to include the families of the returning migrants.

The epidemic might make India's malnutrition crisis worse. The PDS's provisioning of extra grains and certain pulses was primarily intended to address the food security issue that the lockdown especially produced. However, the pandemic has risen the undernutrition and micronutrient malnutrition in India that cannot be solved by only concentrating on the availability of grain. These problems have obstinately persisted in India for many years, mostly as a result of the lack of variety in the population diets. According to the Indian Council of Medical Research (ICMR), Indians often consume more carbs, few proteins, and even few fruits and vegetables than the daily recommended quantities.

In other words, nutritional intake from food categories other than cereal, such as legumes, vegetables, fruits, dairy, and poultry, woefully insufficient and lacking in variety (Hemalatha et al. 2018). The most vulnerable populations are women and children. According to National Family Health Survey 2015–16, stunting affects 38.40 percent of children under the age of five, underweight affects 35.80 percent of children, and anaemia affects 60 percent of children. 53 percent of adult women are anaemic, and 23 percent have a body mass index (BMI) of less than 18.5—indicating insufficient consumption. Their susceptibility is being made worse by supply chain constraints brought on by an epidemic.

Dietary decisions rely on the availability and cost of food (Pingali and Sunder, 2017). According to earlier studies (Gomez et al. 2013; Pinstrup-Andersen 2013), excessive price increases for non-cereal categories reduce their percentage in diet and degrade the quality of total consumption. This chapter examines the possible effects of the pandemic on agricultural commodity prices. The price of a few cereals and non-cereals are examined in this chapter for the year 2019.

Literature Review

Ramesh Chand et al. (2008) had analysed the agriculture sector's volatility in Andhra Pradesh in disaggreagate level. The amount of instability as evaluated by state-level statistics differs significantly from that at the disaggregated level in a big state like Andhra Pradesh. As a result, state-level analysis underestimate agricultural output shocks while overestimate shocks to farm revenue.

Elamin et al. (2011) had made an effort to quantify the degree of instability and the impact of various elements to changes in the mean output of the principal crops planted in the Gezira system. It was discovered that the instability of sorghum, wheat, and cotton had steadily increased across the two periods, whereas the unstable production of groundnut had decreased during the post-liberalization period. It was discovered that all crops' yields and area instability moved in the same manner. To assist attain stability in agricultural production, it was advised that irrigation canals be repaired and that agricultural research and extension be strengthened.

Anjani Kumar et al. (2013) had studied district-level development and instability patterns in agriculture. Within the state and across the entire nation, significant disparities in the production of the agriculture sector were seen. The data analysis emphasised the crucial part that contemporary inputs, such as fertilizers, rainfall, irrigation, labourers, and logistics, play in raising the productivity of the agriculture industry.

Pichad et al. (2014) had studied expansion and volatility of the chickpea production, area, and productivity in the Amravati district. The analysis made use of the coefficient of variation and compound annual growth rate. According to the findings, period II's Compound Annual Growth Rates for area, production, and productivity were all positive and substantial. The coefficient of variance showed that output had the highest levels of instability relative to area and productivity at the overall time, whereas instability in the chickpea region showed less variation overall.

Ayalew et al. (2016) had studied the patterns and geographical disparities in India's maize output. Data from 1980 to 1981 and 2011 to 2012 were examined using decomposition analysis, the Cuddy Della Valle Index, and compound annual growth rate. According to the study, maize output and area both increased throughout that time. The higher yield made this increase possible. According to the study, maize fared better in Andhra Pradesh, Gujarat, Bihar, Madhya Pradesh, Uttar Pradesh and Rajasthan due to the introduction of contemporary maize varieties, which have increased yields and decreased volatility.

Harshita Tewari et al. (2017) had analysed the volatility of wheat area, production, and productivity in Uttar Pradesh. For the years 1990-1991 through 2013-2014, Cuddy Della Valle Index and Decomposition analyses were performed. Bundelkhand area had the lowest productivity, while the western region had the greatest. Additionally, output and productivity were very unstable compared to the area planted with wheat, and the yield impact was discovered to be the primary factor driving production increase.

Based on the above studies, it is found that the past studies have covered single crop. This paper has attempted to cover the major agricultural crops of India during Covid - 19.

Objectives of the study

1. To examine Indicator of Food Price Anomalies and Food Price Inflation in India
2. To study the instability in price of major cereal and non-cereal crops during Covid - 19

Hypothesis of the study

H_1: There has been instability in price of cereal and non-cereal crops during Covid - 19

H_0: There has been no instability in price of cereal and non-cereal crops during Covid - 19

Data and Methodology

Data on Indicator of Food Price Anomalies were collected from FAO with time frame of 2011 to 2020. Food Price Inflation month wise data were collected from FAO during 2019. Data on daily retail price of the principal crops in 2019 were collected from Agmarknet, for analysis averaged to weekly price. The data comprise of three Tier I cities includes Delhi Mumbai and Bengaluru and three Tier II cities includes Bhopal, Guwahati and Lucknow.These cities were chosen because they are the capital cities of various states, making them the perfect representatives of each state. A large geographic range across the nation was also covered by the inclusion of particular cities. It was deliberate to make the distinction between Tier I and Tier II in order to examine any observable variations in food costs between the tiers. The cost of living is generally thought to be higher in Tier I cities than in Tier II cities.

Indicator of Food Price Anomalies

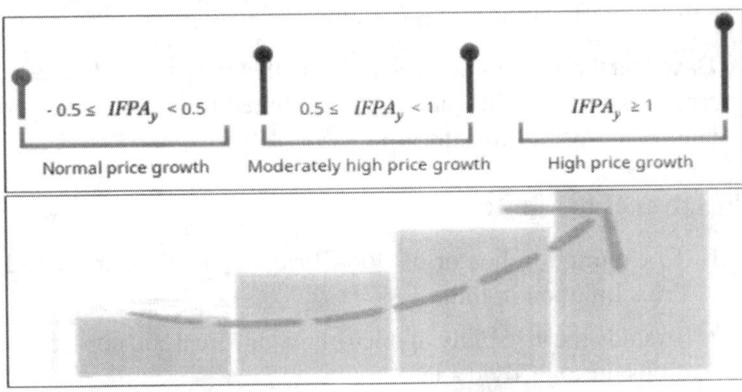

Fig 1. Three levels of IFPA Indicator.

SDG Goal 2 Zero Hunger - Target 2.c.1(adopt actions to maintain the correct operation of food commodity markets and its derivatives and promote prompt access to market information, especially on food reserves) is shown by the Indicator of Food Price Anomalies (IFPA) as it measures of food price volatility and may assist to reduce excessive price volatility. The Indicator of Food

Price Anomalies (IFPA) provides an answer to the question of whether or not a change in prices is abnormal for any given period by analysing growth in food prices over a specific month over a number of years, taking into account seasonality in agricultural markets and inflation.

Price Inflation

Price inflation is a rise in a group of products and services' prices over a certain time frame. Price inflation is brought on by high demand and supply constraints. The rising cost of the production process inputs is another important factor that might contribute to price inflation. A food crisis results from food inflation. Due to a lack of food, the poor have limited access to nutritious food, which results in undernourishment. A one percent increase in food prices causes undernourishment to rise by 0.5 percent.

Cuddy Della Valle Index (CDVI)

The agricultural instability can be measured by different methods, such as the coefficient of variation (CV), dispersion, Cuddy Della Valle Index (CDI), Coppock Instability index, etc. The present study applies the Cuddy Della Valle Index for measuring the instability. Cuddy-Della Valle index attempts to de-trend the CV by using coefficient of determination (R^2). Thus, it is a more accurate way to measure agricultural production instability. This index's value reflects how volatile prices are, and vice versa. Cuddy and Valle (1978) developed the CDVI initially for measuring the trend-driven instability in time series data. The estimable form of the equation is as follows:

$$\text{CDVI} = \text{CV} - \sqrt{1} - \text{AdjR}^2$$

where, **CV** is the coefficient of variation in percent

$\mathbf{R^2}$ is the coefficient of determination from time trend regression adjusted by the number of degrees of freedom.

The CDVI ranges,

Low Instability Index	lies between 0 and 15
Medium Instability Index	lies between 15 and 30
High Instability Index	lies more than 30

Results and Discussion

In this section, Indicator of Food Price Anomalies from 2011 to 2020, Food Price Inflation during 2019 and Price Instability of Principal Crop during 2019.

Indicator of Food Price Anomalies from 2011 to 2020

Year	IFPA	Indicator
2011	-0.11	Normal Price Growth
2012	0.67	Moderately High Price Growth
2013	0.95	Moderately High Price Growth
2014	0.39	Normal Price Growth
2015	0.17	Normal Price Growth
2016	0.36	Normal Price Growth
2017	0.15	Normal Price Growth
2018	-0.59	Normal Price Growth
2019	-0.43	Normal Price Growth
2020	1.05	High Price Growth

Source: FAO-IFPA, applied to the Consumer Food Price Index.

From the above table, it could be inferred over last decades food price showed a normal price growth based on Indicator of Food Price Anomalies (IFPA) after the outbreak of pandemic the food price growth was higher. This price growth challenge will be confronted by amendments of APMC and EC Act in 2020 that will be discussed by the policy reforms section.

Food Price Inflation in India during 2019

Months	Value
January	-1.29
February	-0.07
March	0.66
April	1.38
May	2.03
June	2.37
July	2.33
August	2.96
September	4.70
October	6.93
November	8.73
December	12.16

Source: FAO, 2019.

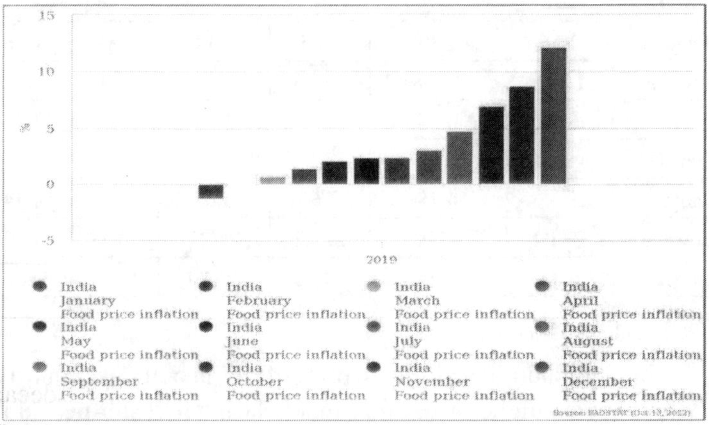

Fig. 2. Food Price Inflation during 2019.

Price inflation is an increase in price of goods and services over a time period. Strong demand and supply shortages tend to cause price inflation. Price inflation can also be caused by the cost of inputs to the production process increasing. Food inflation leads to food crisis. Shortage of food leads to lesser access to high-quality food for the poor, leading to undernourishment among the poor. Food inflation of one percent leads to an increase of 0.5 percent in undernourishment.

During the period 2019, the food price inflation showed a steady increased trend from March to December.

Instability in the price of Principal crops during Covid-19

The instability index of principal crops of India during Covid -19 have been examined and presented using Cuddy Della Valle Index (CDVI). The study period divided into two sub-periods to explain which sub-period influence the overall year price instability represented in the following Tables.

Table 1: Instability in the price of Rice during Covid-19

	1st Half Yearly		2nd Half Yearly		Annual	
	CV	CDVI	CV	CDVI	CV	CDVI
Tier I						
Bengaluru	2.56	1.33	1.43	1.14	2.81	1.46
Mumbai	17.41	15.38	13.45	12.69	16.90	14.34
Delhi	6.30	6.25	2.79	2.77	4.86	4.81
Tier II						
Bhopal	5.25	4.78	3.65	3.65	5.23	4.44
Guwahati	27.04	27.08	25.22	25.27	26.31	26.30
Lucknow	6.38	3.44	13.78	13.70	11.53	11.01

Source: Author's Calculation.

Table 1 explains instability in price of rice in both Tier I and Tier II cities showed higher instability index during first sub-period i.e. from January to June pointed that immediate after the lockdown had increase in the price of rice whereas during the second sub-period i.e., from July to December showed decline in the instability index except Lucknow which showed increase in instability index during second sub-period. So overall during the period, Guwahati falls under medium instability category and remaining all falls under low instability category.

Table 2: Instability in the price of Wheat during Covid-19

	1st Half Yearly		2nd Half Yearly		Annual	
	CV	**CDVI**	**CV**	**CDVI**	**CV**	**CDVI**
Tier I						
Bengaluru	2.58	1.39	4.22	4.13	4.01	3.45
Mumbai	3.51	3.04	9.78	9.73	7.42	7.36
Delhi	5.82	2.10	4.08	2.998	5.64	5.17
Tier II						
Bhopal	4.09	4.04	12.50	12.26	11.12	9.95
Guwahati	0.23	0.15	0	0	0.19	0.12
Lucknow	1.60	0.92	1.17	1.01	1.94	1.27

Source: Author's Calculation.

Table 2 explains instability in price of wheat in both Tier I and Tier II cities revealed the instability index during first sub-period was low but increased during the second sub-period. Though increased trend in instability of wheat price, but still the overall status during the period falls under low instability category.

Table 3: Instability in the price of Redgram during Covid-19

	1st Half Yearly		2nd Half Yearly		Annual	
	CV	**CDVI**	**CV**	**CDVI**	**CV**	**CDVI**
Tier I						
Bengaluru	5.88	3.86	3.67	3.55	6.38	4.64
Mumbai	6.63	1.99	4.06	4.02	9.64	4.42
Delhi	6.76	3.70	3.61	2.55	9.11	3.41
Tier II						
Bhopal	3.67	1.72	10.12	9.16	9.01	6.80
Guwahati	8.51	5.38	11.42	9.62	10.31	8.44
Lucknow	9.09	4.64	2.96	2.96	11.16	5.35

Source: Author's Calculation.

Table 3 explains instability in price of Redgram in both Tier I and Tier II cities by comparing two sub-period revealed that the instability index during first sub-period was low in Mumbai, Bhopal and Guwahati but increased during the second sub-period. Whereas, Bengaluru, Delhi and Lucknow showed higher instability index during first sub-period compared to second sub-period. Though there shown fluctuation in instability index of Redgram price, but still the overall status during the period comes under low instability category.

Table 4: Instability in the price of Blackgram during Covid-19

	1st Half Yearly		**2nd Half Yearly**		**Annual**	
	CV	**CDVI**	**CV**	**CDVI**	**CV**	**CDVI**
Tier I						
Bengaluru	5.25	4.99	14.45	6.30	12.22	9.39
Mumbai	1.90	1.90	10.66	3.84	9.96	6.29
Delhi	3.45	3.46	5.90	4.13	6.52	4.47
Tier II						
Bhopal	17.64	17.44	24.21	24.28	21.41	21.48
Guwahati	8.11	6.39	4.19	2.65	11.43	5.00
Lucknow	4.47	2.37	3.38	2.75	5.41	2.76

Source: Author's Calculation.

Table 4 explains instability in price of Blackgram in both Tier I and Tier II cities had shown increased instability index from first sub-period to second sub-period in all cities except Guwahati. The overall status of instability during the period comes under low instability category except Bhopal which falls under medium instability category.

Table 5 explains instability in price of Greengram in both Tier I and Tier II cities had shown increased instability index from first sub-period to second sub-period in all cities except Mumbai and Guwahati. The overall status of instability during the period comes under low instability category except Bhopal which falls under medium instability category.

Table 5: Instability in the price of Greengram during Covid-19

	1st Half Yearly		2nd Half Yearly		Annual	
	CV	**CDVI**	**CV**	**CDVI**	**CV**	**CDVI**
Tier I						
Bengaluru	5.30	3.43	3.73	3.71	6.00	3.94
Mumbai	4.11	2.88	2.90	1.64	7.37	2.44
Delhi	5.42	4.02	5.01	4.90	5.49	4.79
Tier II						
Bhopal	19.87	16.39	18.10	17.94	19.11	17.93
Guwahati	5.81	4.79	4.50	2.95	5.78	4.40
Lucknow	2.64	2.60	3.25	2.66	4.34	2.74

Source: Author's Calculation

Table 6: Instability in the price of Tomato during Covid-19

	1st Half Yearly		2nd Half Yearly		Annual	
	CV	**CDVI**	**CV**	**CDVI**	**CV**	**CDVI**
Tier I						
Bengaluru	37.79	34.43	33.99	33.68	37.84	37.40
Mumbai	22.50	19.74	34.41	34.65	29.23	29.13
Delhi	16.93	15.07	40.42	39.33	42.73	39.16
Tier II						
Bhopal	21.95	19.39	21.13	21.18	27.94	21.82
Guwahati	45.72	27.43	22.43	20.56	39.93	31.19
Lucknow	43.66	23.10	25.85	23.69	37.17	32.19

Source: Author's Calculation.

Table 6 explains instability in price of Tomato in both Tier I and Tier II cities had shown increased instability index from first sub-period to second sub-period in all cities except Bengaluru and Guwahati. The status of instability during the period comes under medium instability category in Mumbai and Bhopal whereas Bengaluru, Delhi, Guwahati and Lucknow fall under High instability category.

Table 7: Instability in the price of Onion during Covid-19

	1st Half Yearly		2nd Half Yearly		Annual	
	CV	CDVI	CV	CDVI	CV	CDVI
Tier I						
Bengaluru	29.89	23.15	71.02	38.90	106.81	67.55
Mumbai	36.10	17.69	60.12	26.21	94.91	50.22
Delhi	24.49	22.04	46.20	20.66	75.05	41.79
Tier II						
Bhopal	20.97	20.55	53.34	27.72	81.32	50.13
Guwahati	19.78	15.82	42.52	14.10	63.87	33.19
Lucknow	14.50	9.17	41.55	12.47	65.33	32.01

Source: Author's Calculation.

Table 7 explains instability in price of Onion in both Tier I and Tier II cities had shown increased instability index from first sub-period to second sub-period in all cities except Delhi and Guwahati. The overall status of instability during the period comes under high instability category.

Table 8: Summary of the Food Groups in Relative to Price Instability

Food Groups	Commodities	Instability Category
Cereals	Rice	Low
	Wheat	Low
Pulses	Redgram	Low
	Blackgram	Low
	Greengram	Low
Vegetables	Tomato	High
	Onion	High

Source: Author's Calculation.

Conclusion

India was experiencing two extraordinary difficulties in the area of health. To stop the corona virus from spreading, a national lockdown must be enforced first. Then, throughout the lockdown, the smooth supply of essentials must be maintained. Using government data on retail for vital agricultural goods, this study emphasises the impact of price disruption on the nutrition security of Indian population.The study has analysed the price instability using Cuddy Della Valle Index and represented the summary in Table 8. For Cereal crops, the overall status of instability mostly falls under low instability category lies between 0 and 15. For Non – Cereal crops, the overall status of instability also falls under low instability category lies between 0 and 15. For Vegetable crops, the overall status of instability also falls under medium and high instability category lies between 15 and 30. Therefore, the null hypothesis is rejected, there is an instability in the price of the selected crops. The price rise is borne by vegetables. A timely government intervention and pre-existing infrastructure have protected against any supply disruption and stabilised cereal prices.

After the lockdown, prices increased as a result of ineffective attempts to re-establish the supply of fruits, vegetables, and pulses. This over dependence on staple foods will continue to be a significant part of the Indian diet due to the relative stability of grain costs and the higher pricing of other food categories. This will affect consumer expenditure. Because malnutrition is associated with a weakened immune system, it presents a serious issue during the epidemic. According to studies, people with Covid-19 who have immune system damage may have greater death rates. Because of this, vulnerable populations in poorer nations are more susceptible to the virus.The pandemic gives a chance to rethink India's non-staple supply chain and returns the discussion to the need of building immunity through the consumption of fresh, varied, and nutrient-dense diets.

The most impoverished people in India are facing a struggle for obtaining enough food and nutrition. This was due to the looming fears of a reduction in remittance, a loss of financial stability among the migrants and the daily wage employees. Over past ten years, India has achieved significant progress in the fight against malnutrition. To prevent these achievements from being undone, it will be crucial to take early action to ensure the agricultural supply chain for non-staples runs smoothly. For the purpose of ensuring food and nutrition security, this will be crucial.

Policy Reforms and Implications for Food Security during Covid–19

On May 15, 2020, Finance Minister Nirmala Sitharaman revised the Agriculture Produce Marketing Committee Act and the Essential Commodities Act in response to the pandemic in the agricultural sector. The APMC Act permitted the division of the states into distinct market zones. They have their own set of rules and regulations that are developed and put into effect by autonomous APMCs. The APMCs frequently controlled prices in a monopolistic manner, which was harmful to the rural community (Verma, 2020). A change to the statute releases farmers from the requirement to sell at APMCs located inside the states and gives them the freedom to charge competitive prices for their produce.

Essential Commodities Act was a key tool in maintaining particular stock limitations and a price ceiling for essential commodities. However, the act has impeded the establishment of assets in the agricultural sector as well as advantages for farmers. When the crop is abundant, the issue gets worse. Farmers could stabilise the supply and lower price volatility if they were allowed to store extra crops. The EC Act sets harsh penalties on business owners or farmers who are thought to be hoarding. Farmers were forced to sell their goods at throwaway prices in the wholesale markets as a result of the agencies' refusal to invest in storage facilities.

To lessen the negative effects of Covid-19 on the agriculture sector, the central government took the initiative to implement these measures. By amending these laws, the agricultural sector will be able to accomplish two crucial objectives: enhancing farmers' financial stability and promoting farm output diversification by granting them access to improved technologies.

References

Adhikari, A., Narayanan, R., Dhorajiwala, S., &Mundoli, S. (2020). 21 days and counting: Covid-19 lockdown, migrant workers, and the inadequacy of welfare measures in India.

Beniwal, V., & Srivastava, S. (2020). India unveils $22.6 billion stimulus plan to ease virus pain. *Blooomberg Quint*.

Birhanu, A., & I, S. (2016). Trends and regional disparity of maize production in India. *Journal of Development and Agricultural Economics*, 8(9), 193-199.

Chand, R., & Raju, S. S. (2008). Instability in Andhra Pradesh agriculture—A disaggregate analysis. *Agricultural Economics Research Review*, 21(347-2016-16695), 283-288.

Ellis-Petersen, H., & Chaurasia, M. (2020). India racked by greatest exodus since partition due to coronavirus. *The Guardian*, 30.

Gómez, M. I., Barrett, C. B., Raney, T., Pinstrup-Andersen, P., Meerman, J., Croppenstedt, A., & Thompson, B. (2013). Post-green revolution food systems and the triple burden of malnutrition. *Food Policy*, 42, 129-138.

Hemalatha, R., Toteja, G. S., & Bhargava, B. (2018). It's time we turn our attention to quality of food as well. *The Indian Journal of Medical Research*, 148(5), 469.

Kaur, Gurneel. (2020). Coronavirus Lockdown Having an Adverse Effect on Agriculture Sector in India. *Grainmart News*, March 28.https://www.grainmart.in/news/coronavirus-lockdown-having-an-adverse-effect-on-agriculture-sector-in-india/

Kumar, A., & Jain, R. (2013). Growth and instability in agricultural productivity: A district level analysis. *Agricultural Economics Research Review*, 26(347-2016-17113), 31-42.

Moneycontrol. (2020). India's Foodgrain Production Hits Record High 291.95 Million Tonnes for 2019–20 Crop Year. Moneycontrol.com, February 18. https://www.moneycontrol.com/news/business/economy/indias-foodgrain-production-hits-record-high-291-95-million-tonnes-for-2019-20-crop-year-4956711.html

Mahir, M. E. A. E., & Abdelaziz, H. H. (2011). Analysis of agricultural production instability in the Gezira scheme. *Journal of the Saudi Society of Agricultural Sciences*, 10(2), 53-58.

Pingali, P., & Sunder, N. (2017). Transitioning toward nutrition-sensitive food systems in developing countries. *Annual Review of Resource Economics*, 9(1), 439-459.

Pichad, S. P., Wagh, H. J., & Kadam, M. M. (2014). Growth in area, production and productivity of chickpea in Amravati district. *International Research Journal of Agricultural Economics and Statistics*, 5(2), 289-292.

Pinstrup-Andersen, P. (2013). Nutrition-sensitive food systems: from rhetoric to action. *Lancet (London, England)*, 382(9890), 375-376.

Times of India.(2020). India Beats Other Nations in Covid Response: Study. April 11. https://timesofindia.indiatimes.com/india/india-beats-other-nations-in-covid-response study/articleshow/75089755.cms

Tewari, H., Singh, H. P., & Tripathi, U. (2017). Growth and instability in wheat production: A region wise analysis of Uttar Pradesh, India. *International Journal of Current Microbiology and Applied Sciences*, 6(9), 2537-2544.

Tripathi, Rahul. (2020). Govt Issues Fresh Guideline on Exemption for Agriculture Operations.*EconomicTimes*,March27.https://economictimes.indiatimes.com/news/economy agriculture/govt-issues-fresh-guideline-on-exemption-for-agriculture-operations/articleshow/74852088.cms

Verma, Sanju. (2020). Defanging Outmoded APMC Model and Altering Essential Commodities Act Will Unlock Value, Empower Farmer, Writes BJP Economist.Firstpost.com, May 16.https://www.firstpost.com/india/defanging-outmoded-apmc-model-and-altering-essential-commodities-act-will-unlock-value-empower-farmer-writes-bjp economist-8374491.html

4

Trends in Nanotechnology for Sustainable Food Security

N. Swathika

Introduction

As the world population is expected to increase by ten billion people by 2050, there is an alarming challenge to increase the living standard of each individual and achieve sustainability. The United Nations adopted the Sustainable Development Goals in 2017 as an alarming call for all countries, realising that eradicating poverty and other setbacks require simultaneous improvements to the economy, environment, and health sector. To address the sustainable goals, the Waterloo Institute for Nanotechnology (University of Waterloo, Canada) organized a virtual international workshop on nanotechnology for sustainable future in November 2020. The insight of the workshop was to establish the International Network for Sustainable Nanotechnology, whose mission is to bridge the enormous gap between scientific communities, technologists, policy-makers, and governments globally to hasten the adoption of crucially important technology-based solutions to solve existing issues (Pokrajac et al. 2021). We are currently looking into cutting-edge technologies to address the unresolved issues across all industries.

Nanotechnology is one of the most promising technologies in the current century due to its unique size-dependent properties. There is a famous quote "There's plenty of room at the bottom" by Richard Feynman which paved the opening to nanotechnology. Nanomaterials (1-100 nm) possess huge surface area which makes them distinctively different from the bulk materials and find application in various sectors such as agriculture, food, energy, environment, and health. These nanomaterials can be fabricated using top-down and bottom-up approaches. It has recently become clear that nanotechnology is a flexible platform which could offer an effective, affordable, and ecologically acceptable answer to society's significant sustainability issues. To address the problems of food security in a sustainable manner, advanced technologies can be significantly employed in agriculture. Nanotechnology has the potential to enhance food and nutritional security with innovative nanotools for improving crop production and crop protection (Usman et al. 2020). The global market for agricultural nanotechnology is anticipated to expand at a growth rate of 11.7 percent from 2021 to 2028, rising from USD 256 billion (2020) to USD 620.3 billion (2028). Continuous and excessive usage of agricultural inputs such as fertilizers, pesticides etc. cause the alteration in soil microbes, resistance development in pest and pathogens, groundwater contamination and residual toxicity. This repeated pattern leads to economic and sustainable losses. All these problems can be addressed using the nanotechnological interventions. Nanofertilizers, nanopesticides, nanoparticles for seed quality improvement, nano-diagnostics for early detection of plant diseases, etc. draw attention in agriculture research and the advantages of nano-based formulations include sustained, targeted delivery of active molecules, reduced input use efficiency and minimal residual effects (Figure 1). This review emphasises on the areas of nanofertilizers, nanopesticides and nanoparticles for seed quality enhancement.

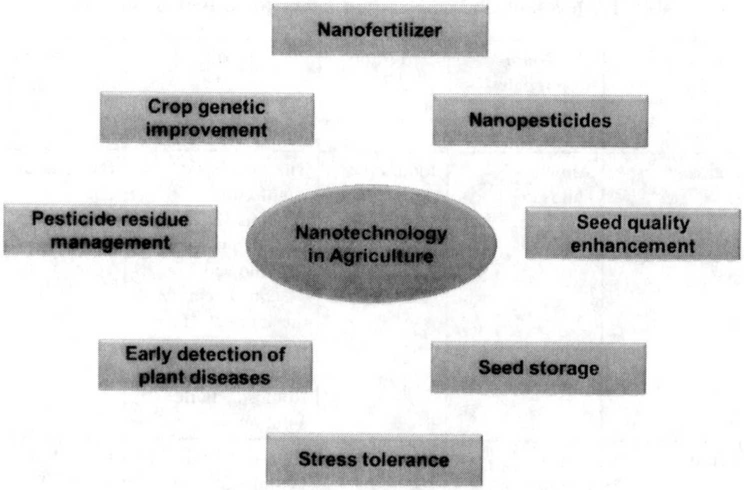

Fig. 1. Application of nanotechnology in Agriculture.

Nanofertilizers—Towards sustainability

It is important to emphasise that the fertilizer usage accounts for one-third of the net crop yield. As a result, fertilizers use must be carried out properly to limit production costs and the ecological harm that results from the excessive consumption of synthetic fertilizers. According to several researchers, nano-fertilizers can act as the greater replacement option for the existing conventional fertilizers because it exhibits three times more nutrient use efficiency than chemical fertilizers due to their greater surface area. Nano-fertilizer is a product with a nanoscale dimension that can be used as an alternative to conventional fertilizer to deliver nutrients. Nanofertilizer can be fabricated through (a) Direct Loading: One or more nutrients can be loaded onto a nano-scale adsorbent. Depending on the surface charge of the adsorbent, surface modification can be performed using suitable surfactants for the anionic and cationic nutrients; (b) Encapsulation of fertilizers using wall materials. Nanofertilizer minimizes the nutrient loss caused by leaching and can enhance nutrient use efficiency (Rahale et al. 2021). Chitosan Nanofertilizer was fabricated using ionic

Table 1. Development of Nanofertilizers for different nutrients

Nano-fertilizer Nutrient	**Nano-particles/ Carrier material**	**Method**	**Results**	**References**
Potassium	Alginate-chitosan	Ionotropic pre-gelation	The developed nanofertilizer contains 29.75 percent potassium and showed sustained release of the nutrient (14.6 percent) after 24 hours in Britton-Robinson buffer solution	(Nido et al. 2019)
Potassium	kappa-carrageenan	Crosslinking method	Controlled release of nutrient was achieved using the potaasium-carrageenan nanofertilizer compared to the conventional fertilizer	(Toledo et al. 2019)
Phosphorus	Hydroxyapatite nanoparticles	-	Productivity of rosemary plants was enhanced using the nanofertilizer	(Elsayed et al. 2022)
Sulfur	Nano Zeolite Surface modification-Hexadecyl trimethyl ammonium bromide	Adsorption	23 percent more surface was absorbed by Surface modified zeolite and recorded controlled release of the nutrient	(Thirunavu-kkarasu & Subram-anian, 2015)
Urea (Nanourea-Modified Hydroxyapatite Nanoparticles)	Hydroxyapatite Nanoparticles	-	The fabricated nanofertilizer was tested on *Abelmoschus esculentus* exhibiting greater nutrient use efficieny & more yields	(Tarafder et al. 2020)

gelation and seed treatment was performed with the developed nanofertilizer which results in 1.6- and 1.7- to 3.0 folds enhanced seedling vigour and reserve food mobilising enzyme activity respectively. Also, it was worth noting that the foliar application of the chitosan nanofertilizer enhanced chlorophyll content, antioxidant activities, greater sucrose translocation and decreased malondialdehyde content (Sharma et al. 2020). Similarly, chitosan-silicon nanofertilizer was synthesised using ionic gelation technique and recorded sustained release of silicon thereby promoting maize growth and yield (Kumaraswamy et al. 2021). Some examples of nanofertilizer of different nutrients are listed in Table 1. IFFCO developed and launched nano urea (20-50 nm) that was approved by the Government of India and the Fertilizer Control Order. Owing to their small size, they can easily enter the stomata pores and other openings causing quicker translocation from source to sink while the unutilized nitrogen retains in the vacuoles and is released slowly.

Nanopesticides

In agriculture, pesticides have become a significant element in minimising the yield losses caused by pests. Numerous environmental and health risks have resulted from the larger dose of chemicals such as herbicides, insecticides, and fungicides in pest management. Nanopesticides refer to the agrochemicals used for plant protection either as nanoparticles directly or in nanocarrier systems that carry the active molecules. NPs including metallic NPs, metal oxide NPs, carbon-based NPs, ceramics, silica, polymeric NPs, clay-based NPs etc. are widely employed in pesticide nanoformulations. Nanopesticides can also be formulated using a suitable nanocarrier system such as nanospheres, nanocapsules, nanosuspensions, nanoemulsions, nanogels etc. The advantages of nanopesticides include enhanced pesticide efficacy, improved stability, minimal pesticide dose requirement, targeted and sustained release of the active ingredients thus overcoming the limitations associated with conventional pesticides (Dangi & Verma, 2021). Various formulations loaded with herbicides (Pontes et al. 2021; Wu et al. 2021), insecticides (Ibrahim et al. 2021; Salama et al.

2022) and fungicides (Bui et al. 2021; Siddhartha et al. 2022) were reported to offer protection against pests (weeds, insects, pathogens). Besides offering plant protection, NPs could also be employed in pesticide residue management (Yan et al. 2020).

Nanotechnology for Seed quality enhancement

Seed Nano-priming- Nano-priming can be used to guard seeds during storage, enhance germination, improve plant growth, and raise the tolerance to biotic and abiotic stresses which can help to minimize the amounts of pesticides and fertilizers needed. Metal NPs, metal oxide NPs, polymer-based NPs and carbon-based NPs have been applied as seed pre-treatment agents for promoting seedling growth, seed germination, and stress tolerance (Elhaj Baddar & Unrine, 2018; Kumar et al. 2018; Maity et al. 2018; Xin et al. 2020). Numerous researches have demonstrated the beneficial effects of NPs in seed applications, particularly to progress seed quality parameters through enhanced H_2O uptake; enzymatic activity and improved defence mechanism (Raja et al. 2019).

Nanofiber technology-Nanofibers can be produced using the electrospinning technique in which the polymer solution is subjected to high voltage (10-30 kV) resulting in the fibers of diameter in nano-regime. Their applications include nanofiber seed coating (Raja et al. 2021), encapsulation of agrochemical inputs (Gao et al. 2021; Kampeerapappun & Phanomkate, 2013), pheromones (Bisotto-De-Oliveira et al. 2015; Kikionis et al. 2017), beneficial microorganisms (De Gregorio et al. 2017; Mukiri et al. 2022) etc.

Seed Storage - During storage, nanoparticles can be one way to maintain viability and vigour by reducing losses caused by biotic and abiotic stress. NPs seed invigoration increases antioxidant activity by quenching the free radicals that cause ageing. Also, the volatile profile of the seeds alters during the ageing process which causes the loss of seed vigour. It was reported that the NPs seed invigoration effectively minimizes the emission of volatile organic compounds during the storage process (Meenakshi et al. 2020).

Conclusion

Nanotechnology has a significant impact on the agriculture sector to enhance crop production and guarantee environmental sustainability to meet future demands. Nanofertilizers have economic advantages over conventional fertilizers such as decreased volatilization and leaching and also can improve crop yield by boosting farmers' profit margin. Nanomaterials or novel nanoformulation of pesticides offers effective management of pest thereby providing safer and environmental friendly crop protection. There are plentiful prospects to utilize nanotechnological strategies to improve seed quality characteristics and to incorporate bioactive molecules and beneficial microbes for smart delivery and improved use efficiency of the active ingredients. Besides nanofertilizers, nanopesticides and nanoparticles for seed quality enhancement, nanotechnology also finds application in the early detection of plant diseases, pesticide residue management, mitigation of biotic and abiotic stresses, and nanotechnology for crop genetic improvement. Before commercialization, the engineered nanomaterials must be studied thoroughly to assess the toxicity potential through biosafety tests. Thus, by employing nanotechnology in agriculture there are ample opportunities to address food security issues and feed every individual by addressing sustainable development goals such as no poverty and zero hunger.

References

Bisotto-De-Oliveira, R., Morais, R. M., Roggia, I., Silva, S. J., SANT'ANA, J., & Pereira, C. N. (2015). Polymers nanofibers as vehicles for the release of the synthetic sex pheromone of Grapholita molesta (Lepidoptera, Tortricidae). *Revista Colombiana de Entomología, 41*(2), 262-269.

Bui, V. C., Nguyen, T. H., Pham, N. T., Vu, H. D., Nguyen, X. C., De Tran, Q., . . . Dai Lam, T. (2021). Curcumin-removed turmeric oleoresin nano-emulsion as a novel botanical fungicide to control anthracnose (Colletotrichum gloeosporioides) in litchi. *Green Processing and Synthesis, 10*(1), 729-741.

Dangi, K., & Verma, A. K. (2021). Efficient & eco-friendly smart nanopesticides: Emerging prospects for agriculture. *Materials Today: Proceedings*, *45*, 3819-3824.

De Gregorio, P. R., Michavila, G., Ricciardi Muller, L., de Souza Borges, C., Pomares, M. F., Saccol de Sá, E. L., . . . Vincent, P. A. (2017). Beneficial rhizobacteria immobilised in nanofibers for potential application as soybean seed bioinoculants. *PLoS One*, *12*(5), e0176930.

Elhaj Baddar, Z., & Unrine, J. M. (2018). Functionalized-ZnO-nanoparticle seed treatments to enhance growth and Zn content of wheat (Triticum aestivum) seedlings. *Journal of agricultural and food chemistry*, *66*(46), 12166-12178.

Elsayed, A. A., Ahmed, E.-G., Taha, Z. K., Farag, H. M., Hussein, M. S., & AbouAitah, K. (2022). Hydroxyapatite nanoparticles as novel nano-fertilizer for production of rosemary plants. *Scientia Horticulturae*, *295*, 110851.

Gao, S., Liu, Y., Jiang, J., Li, X., Ye, F., Fu, Y., & Zhao, L. (2021). Thiram/hydroxypropyl-β-cyclodextrin inclusion complex electrospun nanofibers for a fast dissolving water-bazed drug delivery system. *Colloids and Surfaces B: Biointerfaces*, *201*, 111625.

Ibrahim, S. S., Salem, N. Y., Abd ElNaby, S., & Adel, M. M. (2021). Characterization of Nanoparticles Loaded with Garlic Essential Oil and Their Insecticidal Activity Against Phthorimaea Operculella (Zeller)(PTM)(Lepidoptera: Gelechiidae). *International Journal of Nanoscience and Nanotechnology*, *17*(3), 147-160.

Kampeerapappun, P., & Phanomkate, N. (2013). Slow release fertilizer from core-shell electrospun fibers. *Chiang Mai J. Sci*, *40*(4), 775-782.

Kikionis, S., Ioannou, E., Konstantopoulou, M., & Roussis, V. (2017). Electrospun micro/nanofibers as controlled release systems for pheromones of Bactrocera oleae and Prays oleae. *Journal of chemical ecology*, *43*(3), 254-262.

Kumar, V., Sachdev, D., Pasricha, R., Maheshwari, P. H., & Taneja, N. K. (2018). Zinc-supported multiwalled carbon nanotube nanocomposite: a synergism to micronutrient release and a smart distributor to promote the growth of onion seeds in arid conditions. *ACS applied materials & interfaces*, *10*(43), 36733-36745.

Kumaraswamy, R., Saharan, V., Kumari, S., Choudhary, R. C., Pal, A., Sharma, S. S., . . . Biswas, P. (2021). Chitosan-silicon nanofertilizer to enhance plant growth and yield in maize (Zea mays L.). *Plant Physiology and Biochemistry*, *159*, 53-66.

Maity, A., Natarajan, N., Vijay, D., Srinivasan, R., Pastor, M., & Malaviya, D. R. (2018). Influence of metal nanoparticles (NPs) on germination and yield of oat (Avena sativa) and berseem (Trifolium alexandrinum). *Proceedings of the National Academy of Sciences, India Section B: Biological Sciences*, *88*(2), 595-607.

Meenakshi, G., Raja, K., Renugadevi, J., & Karthikeyan, M. (2020). Inorganic metal oxide nanoparticles seed invigoration for extended storability of sunflower (Helianthus annus) under ambient environment. *Journal of Pharmacognosy and Phytochemistry*, *9*(6), 1302-1306.

Mukiri, C., Raja, K., Senthilkumar, M., Subramanian, K., Govindaraju, K., Pradeep, D., & Ranjan, S. (2022). Immobilization of beneficial microbe Methylobacterium aminovorans in electrospun nanofiber as potential seed coatings for improving germination and growth of groundnut Arachis hypogaea. *Plant Growth Regulation*, *97*(2), 419-427.

Nido, P. J., Migo, V., Maguyon-Detras, M. C., & Alfafara, C. (2019). Process optimization potassium nanofertilizer production via ionotropic pre-gelation using alginate-chitosan carrier. MATEC web of conferences.

Pokrajac, L., Abbas, A., Chrzanowski, W., Dias, G. M., Eggleton, B. J., Maguire, S., . . . Nazar, L. (2021). Nanotechnology for a sustainable future: Addressing global challenges with the international network sustainable nanotechnology. In: ACS Publications.

Pontes, M. S., Antunes, D. R., Oliveira, I. P., Forini, M. M., Santos, J. S., Arruda, G. J., . . . Grillo, R. (2021). Chitosan/tripolyphosphate nanoformulation carrying paraquat: insights on its enhanced herbicidal activity. *Environmental Science: Nano*, *8*(5), 1336-1351.

Rahale, C. S., Subramanian, K., & Lakshmanan, A. (2021). Nanofertilizer in Enhancing the Production Potentials of Crops. *Nanotechnology in Plant Growth Promotion and Protection: Recent Advances and Impacts*, 63-78.

Raja, K., Prabhu, C., Subramanian, K., & Govindaraju, K. (2021). Electrospun polyvinyl alcohol (PVA) nanofibers as carriers for hormones (IAA and GA3) delivery in seed invigoration for enhancing germination and seedling vigor of agricultural crops (groundnut and black gram). *Polymer Bulletin*, *78*(11), 6429-6440.

Raja, K., Sowmya, R., Sudhagar, R., Moorthy, P. S., Govindaraju, K., & Subramanian, K. (2019). Biogenic ZnO and Cu nanoparticles to improve seed germination quality in blackgram (Vigna mungo). *Materials Letters*, *235*, 164-167.

Salama, M. R., Barakat, D. A., & Ibrahim, E.-D. S. (2022). Effectiveness and Persistence of Some Synthetic Insecticides and their Nanoformulation Against Whitefly (Bemisia tabaci) and Aphids (Aphis craccivora) on Fennel Plants and Soil. *Egyptian Journal of Chemistry*.

Sharma, G., Kumar, A., Devi, K. A., Prajapati, D., Bhagat, D., Pal, A., Saharan, V. (2020). Chitosan nanofertilizer to foster source activity in maize. *International journal of biological macromolecules*, *145*, 226-234.

Siddhartha, Verma, A., Bashyal, B. M., Gogoi, R., & Kumar, R. (2022). New nano-fungicides for the management of sheath blight disease (Rhizoctonia solani) in rice. *International Journal of Pest Management*, *68*(3), 217-226.

Tarafder, C., Daizy, M., Alam, M. M., Ali, M. R., Islam, M. J., Islam, R., . . . Khan, M. Z. H. (2020). Formulation of a hybrid nanofertilizer for slow and sustainable release of micronutrients. *ACS Omega*, *5*(37), 23960-23966.

Thirunavukkarasu, M., & Subramanian, K. (2015). Synthesis and characterization of surface modified nano-zeolite fortified with sulphate and its sorption and desorption pattern.

Toledo, M., Migo, V., Alfafara, C., Maguyon-Detras, M. C., & Brutas, C. C. (2019). Process Optimization for the Production of Potassium-Carrageenan Nanofertilizer by Ionic Crosslinking. MATEC Web of Conferences,

Usman, M., Farooq, M., Wakeel, A., Nawaz, A., Cheema, S. A., ur Rehman, H., . . . Sanaullah, M. (2020). Nanotechnology in agriculture: Current status, challenges and future opportunities. *Science of the Total Environment*, *721*, 137778.

Wu, J., Zhai, Y., Monikh, F. A., Arenas-Lago, D., Grillo, R., Vijver, M. G., & Peijnenburg, W. J. (2021). The differences between the effects of a nanoformulation and a conventional form of atrazine to lettuce: physiological responses, defense mechanisms, and nutrient displacement. *Journal of agricultural and food chemistry*, *69*(42), 12527-12540.

Xin, X., Zhao, F., Rho, J. Y., Goodrich, S. L., Sumerlin, B. S., & He, Z. (2020). Use of polymeric nanoparticles to improve seed germination and plant growth under copper stress. *Science of the Total Environment*, *745*, 141055.

Yan, F., Kumar, S., Spyrou, K., Syari'ati, A., De Luca, O., Thomou, E., . . . Rudolf, P. (2020). Highly Efficient Remediation of Chloridazon and Its Metabolites: The Case of Graphene Oxide Nanoplatelets. *ACS ES&T Water*, *1*(1), 157-166.

5

A Study on the Efficacy of Primary Health Centres in Ernakulam District, Kerala

Divya B Kunnath, S. Mobeen Farhana, and
V. Jacquline Raja Bai

Introduction

Health is identified as the state of complete physical, mental and social well-being and not merely the absence of diseases or infirmity (WHO, 1946). Health plays an important role in promoting the well-being of the individual.Any developed healthcare system should be able to deliver the primary, secondary and tertiary services to the public. According to WHO, the main objective of a fully developed healthcare system is to provide better quality and affordable healthcare services to the public. Alma Ata declaration (1978) served as the milestone for recognising primary healthcare as the major factor for the attainment of the goal of health to all. Primary healthcare is defined as "essential care based on practically and scientifically sound, socially acceptable methods and technologically made universally accessible to individuals and families in the community through their full participation and at a cost that community can afford to maintain at every stage of their development and in their spirit of self-reliance and self-determination (WHO, 1978).

The current topic links with the Sustainable Development Goals as well. Health is well placed in SDGs. The health goal (SDG 3) is broad; "Ensure healthy lives and promote well-being for all at all ages". To achieve Universal Health Coverage, it is important that even the poorest section of the society should have access to quality healthcare which is analogous to the main objective of Primary Healthcare as well and that is to provide quality healthcare to rural public.

Primary healthcare constitutes an integral part of any well-developed healthcare system. The primary healthcare identifies the main health issues faced by the public and it provides promotive, curative, preventive and rehabilitative services. Primary healthcare system mainly comprises the Community Healthcare Centres (CHC), Primary Healthcare Centres (PHC) and its sub-centres. The main objective of the primary healthcare system is to provide better quality of healthcare services at an affordable cost to the public. It is unfortunate that the primary healthcare system has many deficiencies in delivering better quality services to the public. The major flaw identified in the primary healthcare system is that even though it is affordable and provides free healthcare services, it is not developed and equipped enough to provide wide range of healthcare services. Thus, the public has to depend on other private hospitals which results in rise in the health expenditure and it becomes difficult for the people especially in rural areas to afford the healthcare.

Primary Health Centres act as an important component of the primary healthcare system to deliver healthcare facilities especially in rural areas. According to the Alma-Ata declaration in 1978, the major principles of the Primary Healthcare centres are accessibility, promotion of health and appropriate technology. Mostly there is only single physician in the PHCs and are usually funded by the government. The various functions of the Primary Healthcare Centres include the provision of the essential drugs, maternal and childcare services, the supply of safe drinking water and provision of basic sanitation facilities, educating the public and conducting health awareness camps etc.

Objectives of the Study

The two main objectives of the study are as follows:

1. To examine the facilities and services of the PHCs from the institutional perspective.
2. To analyse the perception of community in utilising the healthcare services.

Hypothesis of the Study

1. There is no significant association between facilities provided by the PHCs and the area of location of PHCs.
2. There is no significant association between the services provided by PHCs and the location of PHCs.
3. There is no significant association between the number of outpatients in PHCs and their area of location.

Tools of Analysis

The data collected through primary method was analyzed using the SPSS software. Appropriate quantitative and statistical techniques were uzed to analyse the data. Chi square test was performed to see whether there was any association between the facilities provided by the PHCs and the area of location of the PHCs. Correlation was performed to see whether there was an association between the number of outpatients per month in PHCs and the services provided by the PHCs. Multiple regression was done to analyse various factors affecting the number of outpatients in PHCs and Multinominal logistic regression model was performed to examine the factors affecting the PHC visit of the respondents.

Limitations of the Study

The effort to examine the performance of the PHCs in Ernakulam district was subject to the limitation that not all aspects of the healthcare were not analyzed in the study. The study is limited to the geographical boundaries of PHCs in Ernakulam district and data was collected only from 164 respondents due to time constraint.

As the major part of the study was based on the primary data, the memory lapses of the respondents and reluctance from the side of the respondents was another limitation of the study. The time period of the study was another major limitation of the study. The data for the study was collected during the pandemic situation and this was one of the main hindrances for collecting the data from the respondents.

Analysis

Comparison of facilities provided in rural and urban PHCs

The figure below depicts the facilities provided in the rural areas. Considering the infrastructural facilities such as adequate area of the building, designated building, well-furnished building and compound wall as given in the figure, only few PHCs have basic infrastructural facilities in the rural areas. Considering the other facilities only few PHCs in the rural areas have emergency facilities, residential facilities, waiting room facilities, pharmacy, registration counter, separate utilities for men and women and proper waste management system. Rural PHCs did not have operation theatre facilities. The situation prevailing in the rural PHCs was found to be pathetic.

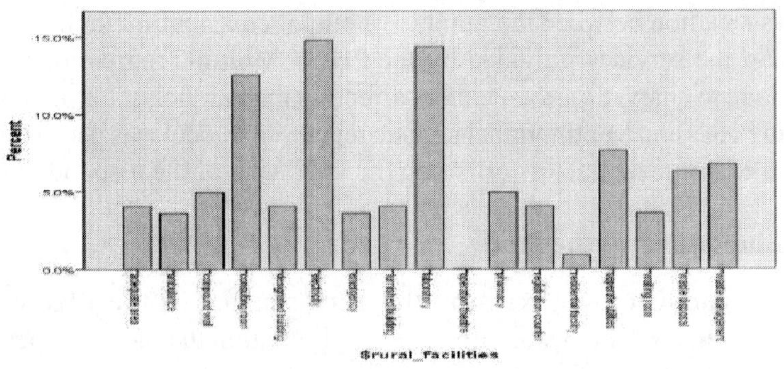

Fig. 1: Facilities provided by the rural PHCs.

The figure 1 depicts the facilities provided by the PHCs in the urban areas. As seen in the figure most of the urban PHCs have proper infrastructural facilities such as compound wall, proper designated building, furnished building and adequate area. As most of the facilities are absent in the rural PHCs, the health workers hesitate to work in such PHCs and it adversely impacts the healthcare needs of the rural public.

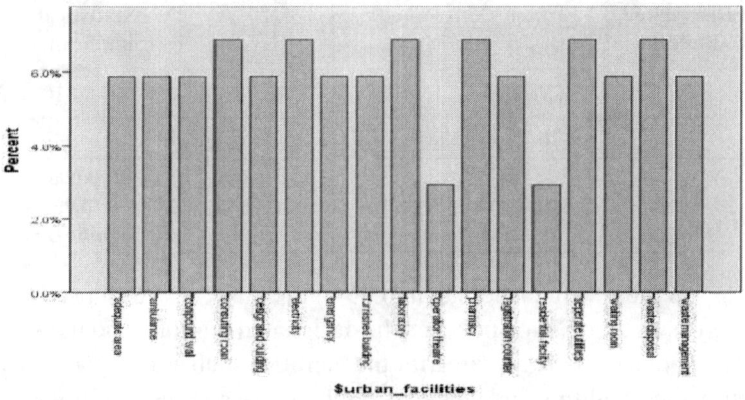

Fig. 2: Facilities provided by urban PHCs.

As seen in the Fig. 2 above most of the urban PHCs have proper infrastructural facilities such as compound wall, proper designated building, furnished building and proper adequate area. Unlike the rural PHCs few urban PHCs have operation theatre facilities. Majority of the urban PHCs have other facilities such as consulting room, emergency, laboratory facilities, registration counter, residential facilities, separate utilities, waiting room and proper waste management system. Compared to the rural PHCs most of the urban PHCs have all the facilities.

Association between the healthcare facilities and area

To find out if there was any association between the facilities provided by the PHCs and the area in which they are located Chi square test was performed. In total 41 PHCs were surveyed, out of which 34 are rural PHCs and seven of them are Urban PHCs. The huge disparity in number of rural and urban PHCs surveyed is due to the fact that most of the PHCs are residing in rural areas and the main objective of the PHCs is to meet the healthcare needs of the rural people.

The results of the Chi square test are as follows:

Table 1: Chi square test

Designated Building	Area		Total	Statistical Significance
	Rural	Urban		
No	25(73.5%)	1(14.3%)	26(63.4%)	X^2=8.810
Yes	9(26.5%)	6(85.7%)	15(36.6%)	Df=1
Total	**34(100%)**	**7(100%)**	**41(100%)**	**P=0.003 P<0.05 Significant**

As the p value is less than 0.05 in the Table 1, we can see that there is an association between the designated building and the area. The above statistics is same for the facilities such as adequate area, furnished building and well plastered floor as all these facilities are related. It has been observed that majority of the rural PHCs do not have a proper designated building and other infrastructural facilities such as adequate area, proper furnished building, and well plastered floor. This have been identified as the prime factor that contributes to the understaffing in rural PHCs. The health workers find themselves hesitant to work in rural PHCs due to non-availability of resources and thus making it difficult for the rural public to have a proper access to quality healthcare services. In contrary to this most of the urban PHCs have proper designated building.

The Table 2 shows that 97.1 percent of the rural PHCs have proper electricity and only 2.9 percent of the rural PHCs does not have proper electricity. Out of the 7 urban PHCs all the PHCs have proper electricity. As shown in the table, value is equal to 0.538 and thus there is no significant association between the electricity and the area of PHCs. It has been observed that irrespective of the location of PHCs most of the PHCs have proper electricity facilities.

Table 2: Association between the provision of electricity in rural PHCs and its area of location

Electricity	Area		Total	Statistical Significance
	Rural	Urban		
No	1(2.9%)	0(0%)	1(2.4%)	X^2=0.380
Yes	33(97.1%)	7(100%)	40(97.6%)	Df=1
Total	**34(100%)**	**7(100%)**	**41(100%)**	**P=0.538 P>0.05 Not Significant**

Table 3: Association between provision of emergency facilities in PHC and its area of location

Electricity	Area		Total	Statistical Significance
	Rural	Urban		
No	26(76.5%)	1(14.3%)	27(65.9%)	X^2=9.802
Yes	8(23.5%)	6(85.7%)	6(57.28%)	Df=1
Total	**34(100%)**	**7(100%)**	**41(100%)**	**P=0.002 P<0.05 Significant**

Table 3 shows that out of the total 34 rural PHCs, 76.5 percent did not have an emergency facility and only 23.5 percent have the emergency facility. Out of the total seven urban PHCs 14.3 percent didn't have an emergency facility but the rest 85.7 percent have emergency facility. This has been identified as the prime reason for the lack of services in the rural PHCs. Without proper emergency backup, the medical staff find it difficult to conduct services in the rural PHCs. P value is 0.002 which is less than 0.05 and hence there is significant association between the emergency backup facility and the area of PHC.

Area-wise comparison of services provided in the PHCs

The services provided by both rural and urban PHCs were analysed separately. In total 34 PHCs were rural PHCs and the rest seven were urban PHCs.

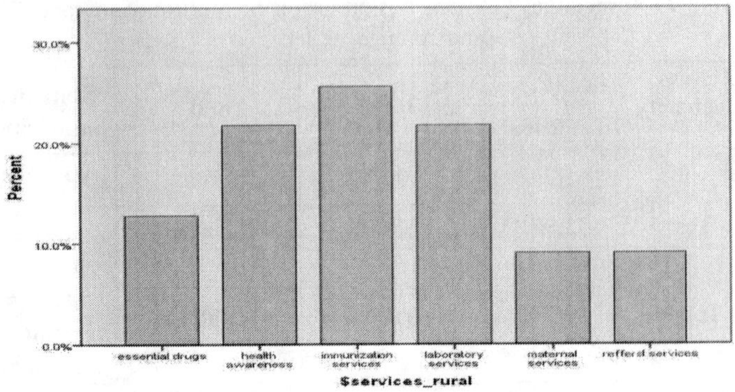

Fig. 3: Services provided by the Rural PHCs.

The Figure 3 represents the services provided by the rural PHCs. Among the various services provided by the rural PHCs; immunization services, laboratory services and health awareness classes were found to be in largest proportion. On the other hand, only few rural PHCs provided maternal and referral services. This is due to the fact that most of the rural PHCs lack basic infrastructural facilities and other facilities such as operation theatres, bed screen, ambulance etc. Another important thing to be noted here is that most of the rural PHCs are in deficit of the essential drugs. Due to lack of proper infrastructural facilities most of the PHCs situated in the rural areas did not have a pharmacy and thus were devoid of essential drugs.

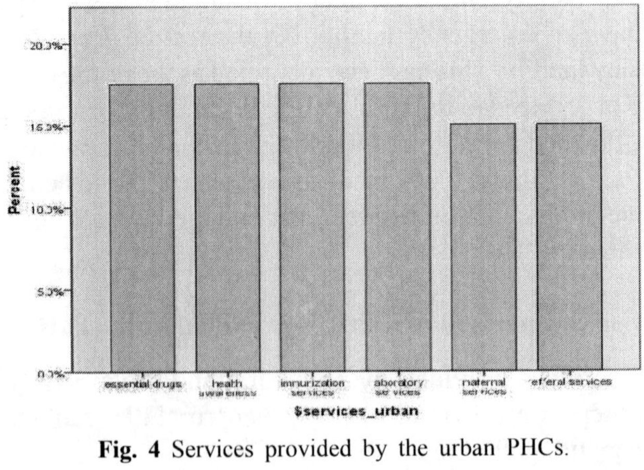

Fig. 4 Services provided by the urban PHCs.

The Figure 4 depicts the services provided by the PHCs in urban areas. As clearly seen from the figure, among the various services provided, provision of essential drugs, conducting health awareness camps, immunization services and laboratory services showed a higher percentage. It was also observed that unlike the PHCs located in the rural areas, most of the PHCs in urban areas provided maternal and referral services. This was due to the fact that most of the PHCs located in the urban areas were not devoid of the facilities.

Table 4: Descriptive statistics of number of outpatient services per month in rural PHCs

	N	Minimum	Maximum	Mean	Std. Deviation
Number of Outpatient services in a month	34 34	250	3700	1561.76	1015.640

The Table 4 indicates that mean number of outpatients per month in rural PHCs is 1561.76 and the next table shows that mean number of outpatients per month in urban PHCs is 878.5. It has been observed that even though urban PHCs provide more referral and maternal services compared to the rural PHCs, the number of outpatients is higher in the rural PHCs than in the urban PHCs. The main reason identified behind this fact is that most of the people residing in the rural areas were found to have poor income and even though the rural PHCs didn't have proper infrastructural facilities and didn't provide many services compared to the urban PHCs.

Table 5: Descriptive statistics of number of outpatients per month in urban PHCs

	N	Minimum	Maximum	Mean	Std. Deviation
Number of Outpatient services in a month	7 7	250	1700	878.57	575.078

In the urban areas even though more services were provided, they are only few in number and most of the people in urban areas have higher income and they depend more on the private hospitals for their health issue.

Correlation between number of referral services per month and number of outpatients in rural PHCs

The table below, represents the correlation between the number of outpatient services per month and number of referral services per month. The sig value is equal to 0.00 which is less than 0.05 and thus shows correlation between the number of referral services and number of outpatients per month, even though only few PHCs in rural areas provide referral services we can see a positive correlation between the referral services and number of outpatients in rural PHCs. Thus, most of the services provided in the rural PHCs and number of outpatients are correlated. This shows that most of the rural public depends on the services provided by the rural PHCs even though most of these PHCs lack basic facilities.

Table 6: Correlation between number of referral services per month and number of outpatients in rural PHCs

		Number of outpatient services in a month	**Referral services number**
Number of outpatients services in a month	Pearson	1	.897**
	Correlation		
	Sig. (2-tailed)		.000
	N	34	34
Referral services number	Pearson	.897**	1
	Correlation		
	Sig. (2-tailed)	.000	
	N	34	34

** Correlation is significant at the 0.01 level (2-tailed).

Table 7: Correlation between number of immunization services per month and number of outpatients in urban PHCs

		Number of outpatient services in a month	**Referral services number**
Immunization Number in a month	Pearson	1	.932**
	Correlation		
	Sig. (2-tailed)		.002
	N	7	7
Number of outpatient services in a month	Pearson	.932**	1
	Correlation		
		0.002	
	Sig. (2-tailed)	7	7
	N		

** Correlation is significant at the 0.01 level (2-tailed).

The Table 7 represents the correlation between the number of immunization services and number of outpatients per month in the rural PHCs. From the table we can see that the sig value is equal to 0.000 which is less than 0.05. Thus, we can say that there is correlation between the immunization services and number of outpatients per month. Among the services provided by both rural and urban PHCs, immunization services show a greater percentage. Even though rural PHCs lacks basic facilities, most of the people in rural areas visit rural PHCs to avail immunization services.

Table 8 shows the correlation between the number of outpatient services per month and number of health awareness classes conducted per month. The sig value is 0.197 which is greater than 0.05 and thus there is no correlation between the number of outpatients and number of health awareness classes conducted per month in the urban PHCs which means that even if the number of health awareness classes per month is increased it will not increase the number of outpatients per month in the urban PHCs.

Table 8: Correlation between number of health awareness classes per month and number of outpatients in urban PHCs

	Number of outpatient services in a month	**Health Awareness number**
Number of outpatient Pearson services in a month Correlation Sig. (2-tailed) N	1 7	.554 .197 7
Health Awareness number Pearson Correlation Sig. (2-tailed) N	.554 .197 7	1 7

Multiple regression for factors affecting the number of outpatients

Factors affecting the number of outpatients per month: number of immunization services per month, number of laboratory services per month, number of health awareness classes conducted, availability of operation theatre, ambulance availability, sufficient staff and referral services.

$$Yi= \beta_{0\,+}\beta_1X_{1i}+\beta_2X_{2i+}\beta_3X_{3i}+\beta_4X_{4i}+\beta_5X_{5i}+\beta_6X_{6i}+\beta_7X_{7i}+u_i$$

Number of outpatients per month=β_0+β_1(number of immunization services per month) +β_2(number of laboratory services per month) +β_3(number of health awareness classes conducted per month) +β_4(operation theatre facilities) +β_5 (ambulance facilities) +β_6(sufficient staff) +β_7(referral services)

Table 9 shows the R square and adjusted R square values are given in the model summary. The R^2 value is 0.909 and adjusted R^2 value is 0.889 which means that 88.9 percent of variance in the dependent variable is explained by the independent variables. Now moving down to the ANOVA table, we can see that model is statistically significant as the p value is 0.00 which is less than

0.05. Moving on to the p values of the independent variables we can see that number of immunization services per month, laboratory services per month and sufficient staff facility has p value less than 0.05 and thus we can say that the number of outpatient services will increase when the number of immunization services per month, number of laboratory services per month and when there is sufficient staff facilities in the PHCs.The p values of the other independent variables is less than greater than 0.05.

Table 9: Model Summary

Model	R	R Square	Adjusted Square R	Std Error of the estimate
1	.953a	.909	.889	327.605

Table 10: Coefficients

Model		Unstandardized Coefficients		Standardized Coefficients		
		B	Std. Error	Beta	t	Sig.
1	(Constant)	-218.143	118.839		-1.836	.075
	immunization number	1.422	.199	.645	7.161	.000
	laboratory services number	.478	.211	.181	2.263	.030
	health awareness number	8.906	9.071	.106	.982	.333
	operation theatre	-311.113	223.765	-.083	-1.390	.174
	ambulance	-270.507	209.840	-.132	-1.289	.206
	staff sufficient	444.814	168.363	.225	2.642	.013
	referral services	253.135	166.462	.129	1.521	.138

a. Dependent Variable: Number of outpatient services in a month

Classification of respondents on the basis of area:

The Table 11 shows the classification of respondents on the basis of area. Out of the total 164 respondents surveyed for the study, majority of the respondents 57.3 percent were from the rural areas and 42.7 percent of the respondents were from the urban areas.

Table 11: Area of the respondents

Area of the respondents	Frequency	Percent
Rural	94	57.3
Urban	70	42.7
Total	**164**	**100.0**

Out of the total 164 respondents surveyed for the study, majority of the respondents, 57.3 percent were from the rural areas and 42.7 percent of the respondents were from the urban areas.

Table 12: Classification of respondents based on Gender

Gender of the respondents	Frequency	Percent
Male	94	57.3
Female	70	42.7
Total	**164**	**100.0**

In total 164 respondents were surveyed for the study. Out of which 94(57.3%) of respondents are male and 42.7 percent of the respondents are female.

Table 13: Whether the respondents have ever visited PHC

Area of Respondents	Whether respondents have ever visited PHC		Total
	No	Yes	
Rural Count % within ever visited PHC	18 26.1%	76 80.0%	94 57.3%
Urban Count % within ever visited PHC	51 73.9%	19 20.0%	70 42.7%
Total Count % within ever visited PHC	69 100.0%	95 100.0%	164 100.0%

From Table 13, it is clear that, out of the total 164 respondents, only 95 respondents have visited PHC and 69 respondents have never visited PHC. Out of the total 94 rural respondents, 80 percent of the respondents have visited PHC and 18 percent of the respondents have never visited PHC and out of the total 70 urban respondents surveyed for the study, only 20 percent of the respondents have visited PHC and 73.9 percent of respondents have never visited PHC. Thus, we can clearly see that majority of the respondents who have visited PHC are from the rural areas and only a few percentages of respondents from urban areas have visited PHCs. The major reason identified for the above trend was that most of the people from rural areas belonged to the BPL category and they couldn't afford high healthcare expenditures while the respondents from the urban areas were able to avail services from the other healthcare providers.

Table 14: How often rural respondents visit PHC

How often Rural respondents visit PHC	Frequency	Percent
Regularly	48	63.2
Once	8	10.5
Occasionally	20	26.3
Total	76	100.0

Table 14 represents how often rural respondents visit PHCs. The respondent visit was categorized as visited regularly, visited once and visited occasionally. Out of the total 76 rural respondents who have visited PHCs, majority of the respondents (63.2%) have visited PHCs regularly, 26.3 percent of respondents have visited occasionally and only 10.5 percent have visited once. Majority of the rural respondents do visit PHCs regularly as the services provided in the PHCs are free of cost.

Table 15: How often urban respondents visit PHC

How often Urban respondents visit PHC	Frequency	Percent
Regularly	4	21.1
Once	9	47.4
Occasionally	6	31.6
Total	19	100.0

Table below depicts how often the urban respondents visit PHCs. Out of the total 19 respondents who have visited PHC, majority of the respondents (47.4%) have visited PHC only once while 31.6 percent of respondents visits PHCs occasionally and 21.1 percent of respondents visits regularly. Most of the urban respondents have visited PHC only to avail the pandemic services. The present covid situation have compelled most of the respondents to visit PHCs to avail pandemic services such as covid vaccination, community awareness classes etc.

Factors affecting PHC visit of the respondents

A multinomial logistic regression (MLR) is taken up to recognise the factors affecting PHC visit of the respondents. The factors affecting PHC visit of the respondents have been recognized as: good quality of services, nearness to home, convenient working hours, qualified doctors, waiting time, lack of transport, distance, facilities, poor quality of service, awareness programmes, service pandemic, and status symbol.

As we have three categories for the outcome variable, we have two regressions for this analysis:

$$\begin{aligned} ln\left(\frac{p(HowOften = Regularly)}{p(HowOften = Once)}\right) \\ &= b_{10} + b_{11}(GoodQualService = 0) + b_{12}(GoodQualService = 1) \\ &+ b_{13}(NearHome = 0) + b_{14}(NearHome = 1) \\ &+ b_{15}(ConvenientHours = 0) + b_{16}(ConvenientHours = 1) \\ &+ b_{17}(AwarenessProgram = 0) + b_{18}(AwarenessProgram = 1) \\ &+ b_{19}(LessWaitingTime = 0) + b_{20}(LessWaitingTime = 1) \\ &+ b_{21}(ImmunizServices = 0) + b_{22}(ImmunizServices = 1) \\ &+ b_{23}(GoodDealStaff = 0) + b_{24}(GoodDealStaff = 1) \\ &+ b_{25}(EasyMedicine = 0) + b_{26}(EasyMedicine = 1) \\ &+ b_{27}(Servicespamdemic = 0) + b_{28}(Servicespamdemic = 1) \\ &+ b_{29}(LowCost = 0) + b_{30}(LowCost = 0) \end{aligned}$$

Regression equation (2):

$$\begin{aligned} ln\left(\frac{p(HowOften = occassionally)}{p(HowOften = once)}\right) \\ &= b_{40} + b_{41}(GoodQualService = 0) + b_{42}(GoodQualService = 1) \\ &+ b_{43}(NearHome = 0) + b_{44}(NearHome = 1) \\ &+ b_{45}(ConvenientHours = 0) + b_{46}(ConvenientHours = 1) \\ &+ b_{47}(AwarenessProgram = 0) + b_{48}(AwarenessProgram = 1) \\ &+ b_{49}(LessWaitingTime = 0) + b_{50}(LessWaitingTime = 1) \\ &+ b_{51}(ImmunizServices = 0) + b_{52}(ImmunizServices = 1) \\ &+ b_{53}(GoodDealStaff = 0) + b_{54}(GoodDealStaff = 1) \\ &+ b_{55}(EasyMedicine = 0) + b_{56}(EasyMedicine = 1) \\ &+ b_{57}(Servicespamdemic = 0) + b_{58}(Servicespamdemic = 1) \\ &+ b_{59}(LowCost = 0) + b_{60}(LowCost = 1) \end{aligned}$$

In the first equation intercept, good quality of services, pandemic services, health awareness camps, immunization services and convenient hours are found to be significant. Here all the variables are categorical. As MLR is a likelihood probability function, we have the base response as "once".

The results suggest that the odds of respondents visiting PHC regularly with respect to once, increases 16.2 times when the quality of services are increased by one unit which signifies that the respondents are more likely to visit PHC when the quality of services increases. Similarly, one unit of increase in pandemic service makes the respondent 2.4 times more likely to visit PHC regularly than once. At the same time when the number of health awareness classes are increased by one the respondents are 3.2 times more prone to visit PHC regularly with respect to visiting PHC once. Immunization services are also found to be a significant factor influencing the respondent's visit to PHC. When the immunization services are increased by one unit, the respondents are four times more likely to visit PHC regularly than once. Convenient hours of functioning of PHC also influences the respondents visit to PHC. There are both part time and full time PHCs. Most of the respondents who are working finds working hours of part time PHCs inconvenient. When the working time of PHCs are increased by one hour the respondents are 3.1 times more likely to visit PHC regularly with respect to once. Results of the second equation of the MLR suggests that intercept, immunization services and health awareness classes are found to be significant. With a one unit of increase in the immunization services the respondents are two times more likely to visit PHC occasionally than once. When the number of health awareness camps are increased by one, the respondents are 1.1 times more likely to visit PHC occasionally than once.

Reasons for rural respondents opting visit to PHC

The majority of the respondents have opted PHC due to the low cost of services. Other major reasons why the rural respondents opt visit to PHC are due to the convenient hours of functioning of PHC and small distance from the residence to PHC. As shown in the figure only few respondents have opted for PHC for the reasons such as easy availability of medicines and good quality of services.

Reasons for urban respondents opting visit to PHC

The various reasons are convenient hours, easy availability

of medicines, good quality of services, good dealing of staff, less waiting time, low cost of services, near to home and qualified doctors. Similarly, to rural PHCs most of the urban respondents also choose PHC due to low cost of services. An important thing to be noted is that unlike rural PHCs, a good proportion of urban respondents opt PHCs due to easy availability of medicines, good quality of services and qualified doctors. Only few urban respondents opt PHC as it is near to home.

Conclusion

Most of the rural PHCs are devoid of the basic infrastructural and other facilities compared to the urban PHCs. It is important to note that only few rural PHCs have maternal and referral facilities and no rural PHCs have inpatient facilities and this prevents the rural public from visiting the PHCs during their emergencies. Most of the rural PHCs are also deficit of medical staff and do not have proper availability of medicines. While the number of outpatients per month in the PHCs are taken into consideration, we can see that rural PHCs have a greater number of outpatients than the urban PHCs. This is mainly because even though the services and facilities of rural PHCs are not good they are forced to visit such PHCs to reduce their healthcare expenditures. On the other hand, even though the services and facilities of urban PHCs are of better quality the number of outpatients per month is low as most of the urban households considers visiting the urban PHCs as low prestige. Majority of the rural respondents find the services of PHCs ineffective and are not satisfied. Most of the urban respondents are not even aware about the location and services being provided.

References

Anitha, C. V., & Thimmaiah, N. (2013). Accessibility of Primary Healthcare Services: A Case Study of Kadakola Primary Health Centre in Mysore District.

Dar, K. (2015). Utilization of the services of the Primary Health Centres in India–an empirical study. *Journal of Health, Medicine and Nursing*, *16*, 56-59.

Khursheed, H. (2017). Empirical analysis of determinants of patient satisfaction: A case study of primary health centres. *Journal of Global Economics*, *4*(234), 2.

Lahariya, C., Sundararaman, T., Ved, R. R., Adithyan, G. S., De Graeve, H., Jhalani, M., &Bekedam, H. (2020). What makes primary healthcare facilities functional, and increases the utilization? Learnings from 12 case studies. *Journal of family medicine and primary care*, *9*(2), 539.

Purohit, B., Patel, D., & Purohit, S. (2014). A study of organizational values in government run primary health centres in India. *Journal of Health Management*, *16*(2), 303-313.

Rajpurohit, A. C., Srivastava, A. K., & Srivastava, V. K. (2013). Utilization of primary health centre services amongst rural population of northern India-some socio-demographic correlates. *Indian Journal of Community Health*, *25*(4), 445-450.

Varatharajan, D, Thankappan, R and Sabeena, Jayapalan. (2004). Assessing the performance of primary health centre under decentralised government in Kerala, India. *Health policy and planning*, 19(1) :14-51.

District Medical Office (Health), Kerala http://dmohekm.in

District Medical Office (Health), Ernakulam https://ernakulam.nic.in

District Medical Office (Health), Kerala http://dmohekm.in

District Medical Office (Health), Ernakulam https://ernakulam.nic.in

6

A Sustainable Medical Practices in Treating Life Style Disorders

Zakiyyah Taj and *C. J. Lakshmi*

Introduction

Healthcare refers "to the prevention, diagnosis, treatment, amelioration, or cure of disease, illness, injury, and other physical and mental problems in people". Access to healthcare differs by country, community, and individual, and is influenced by social and economic variables as well as healthcare policies. TRM is not only becoming a progressively larger component of the trillion-dollar global health, wellness, beauty, health tourism, and pharmaceutical businesses, but it also continues to be a common source of preventative, promotional, and economical curative health for billions of people who may have little contact with healthcare services, especially beyond the primary level.

1.1 Meaning of Traditional alternative and Allopathy medicine

Traditional alternative medicine: "any of a range of medical therapies that are not regarded as orthodox by the medical profession, such as herbalism, naturopathy, and crystal healing".

Allopathy medicine: "relating to or being a system of medicine that aims to combat disease by using remedies (such as drugs or surgery) which produce effects that are different from or incompatible with those of the disease being treated".

1.2 History of Traditional Alternative Medicine

Traditional medicine is an amalgamation of information, skill, and practices that are based on native cultural theories, beliefs, and experiences, whether they can be explained. It is used for treatment, restorative care, prevention, diagnosis, and maintenance of the physical and psychological state.

Traditional medical practices in India are well-known. Even the ancient Vedas and other writings refer to medical systems. When ancient scriptures like the "Rigveda" (1700-1100 BCE), "Yajurveda 400-1000 BCE", and "Atharva Veda" (1400-1000 BCE) chronicled healthcare demands and diseases, traditional medicine and its healthcare records date back to 5000 BCE (1200 BCE). The usage of herbs and polyherbal mixtures of later writings, such as the "Charaka Samhita" (990 BCE), "Sushruta Samhita" (660 BCE), and "Dhanvantari Nighantu" (1800 CE).

In Indian traditional medicine, the application of drugs that are said to be indigenous to India or to have arrived there from another country and absorbed into Indian culture. India's only recognized traditional medical systems are Ayurveda, Siddha, Unani, Yoga, Naturopathy, and Homeopathy. Even though homeopathy was introduced to India in the eighteenth century, it has been fully assimilated into Indian culture and has developed into a traditional system as a result, which has allowed it to be improved and integrated into Indian traditional medicine.

1.3 History of Allopathy

The Greek words "allos" and "pathos," which both imply the opposite and "to suffer," are where the word "allopathic" originates. In the 1800s, German physician Friedrich Samuel Hahnemann created this term.

Compared to today's doctors, those in the early 1800s had a lot less to offer patients, yet they were nevertheless able to do a lot more because of their knowledge of anatomy, the natural history of many illnesses, and early insights into the microbiological genesis of infectious diseases.With time, traditional medicine started to develop and implement innovations including vaccines, insulin, new medications, surgical procedures, cancer treatments, and public health campaigns that not only efficiently treated symptoms but also eliminated or prevented a large number of diseases.

Objectives of the study

The objectives of the study are to evaluate the socio-economic factors that influence the choice of Traditional alternative medicine and conventional medicine. To analyse the cost and effectiveness involved in both practices in treating lifestyle disorders. To assess the funds released by the government towards Traditional alternative medicines and Allopathy.

Data and methodology

This research is based on both primary and secondary data sources. The questionnaire was used to obtain primary data. This data was collected in Karnataka, Bengaluru. The sample was selected through a non-probability snowball distribution sample technique and the sample size is 70 for each form of medicine.The tools used for primary data analysis were Cost-effective ratio and ANOVA to understand the cost-effective medicine, Chi-Square was used to see the effect of socioeconomic factors, and Pie charts to understand the preferences of people. Secondary data is gathered from the Ministry of Health and Family Welfare and the Ministry of AYUSH's websites.The acquired data were analyzed using a trendline.

Hypothesis

1. Null hypothesis (H_0): There is no significant difference between socio-economic factors (age, gender, income, occupation, and education qualification) and choice of medicine.

2. Alternative hypothesis (H_1): There is a significant difference between socio-economic factors (age, education qualification, income, occupation, and gender) and choice of medicine.

Null hypothesis (H_0)- states there is no significant impact on cost and effectiveness between alternative and allopathic care in treating lifestyle disorders.

Alternative hypothesis (H_1)- states there is a significant impact on cost and effectiveness between alternative and allopathic care in treating lifestyle disorders.

Results

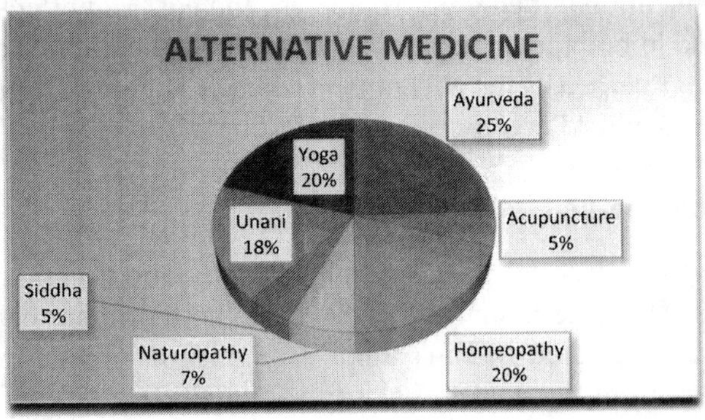

Fig. 1: Preference for forms of Alternative medicine.

The Figure 1 shows that most people use Ayurveda (25%) followed by Homeopathy (20%) and Yoga (20%) to treat lifestyle disorders.

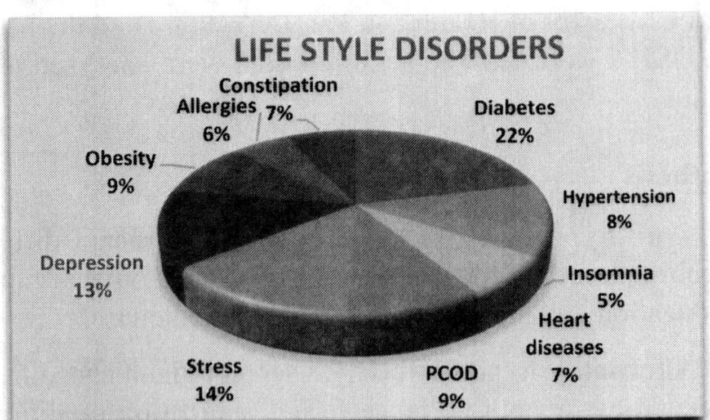

Fig. 2: Show the percentage of people who opted for alternative medicine to treat lifestyle disorders.

The above figure depicts the percentage of patients who opted for alternative medicine to treat lifestyle disorders, 22 percent for diabetes, 14 percent for stress, 13 percent for depression, 9 percent for obesity, 9 percent for PCOD, 8 percent for hypertension, 7 percent for heart diseases, 7 percent for constipation, 6 percent for allergies, and 5 percent for insomnia.

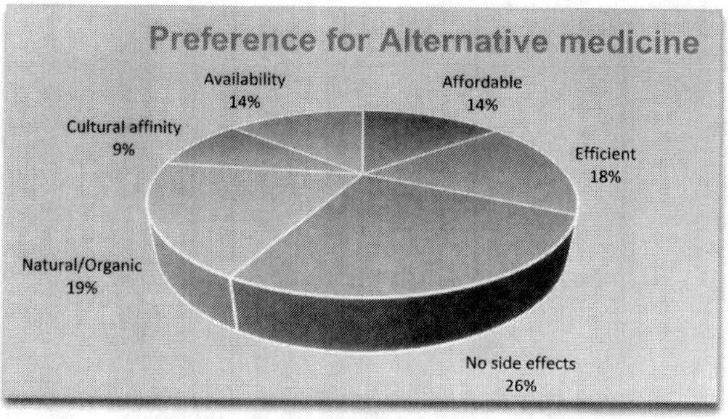

Fig. 3: Show the preferences of people for using alternative medicine.

People's preferences for alternative medicine are depicted in the figure above. Alternative medicine is used by people for a variety of reasons, including no side effects (26%), natural/organic (19%), efficient (18%), affordable (14%), availability (14%), and cultural affinity (9%).

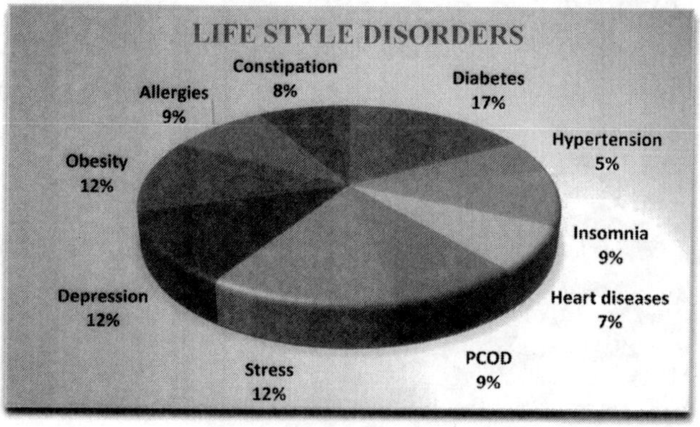

Fig. 4: Show the percentage of people who opted for allopathy medicine to treat lifestyle disorder.

The above figure depicts the percentage of patients who opted for allopathy medicine to treat lifestyle disorders, 17 percent for diabetes, 12 percent for stress, 12 percent for depression, 12 percent for obesity, 5 percent for hypertension, 7 percent for heart diseases, 8 percent for constipation, 9 percent for allergies, 9 percent for PCOD, and 9 percent for insomnia.

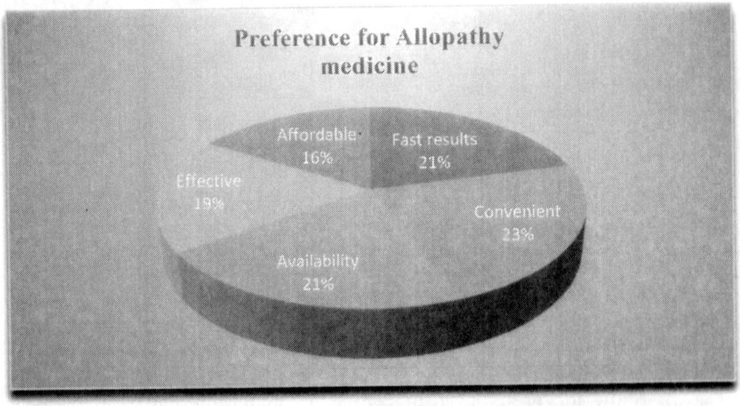

Fig. 5: Show the preferences of people for using allopathy medicine.

People's preferences for allopathy medicine are depicted in the figure above. Allopathy medicine is used by people for a variety of reasons, including convenience (23%), availability (21%), fast results (21%), efficiency (19%), and affordability (16%).

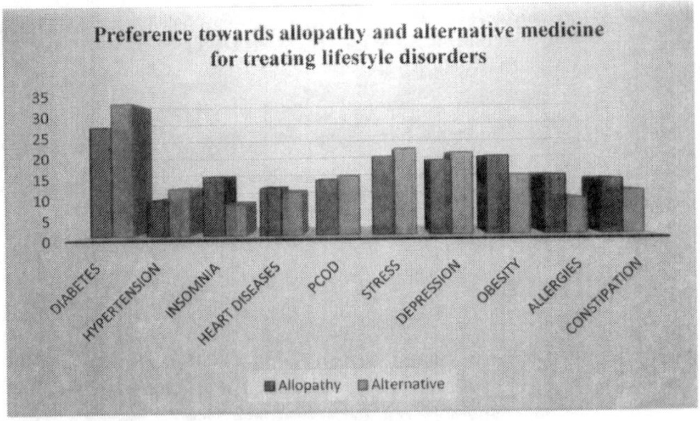

Fig. 6: Preferences toward allopathy and alternative medicine for treating lifestyle disorders.

The above figure shows the preferences of people for allopathy and alternative medicine to treat lifestyle disorders. We can depict that diabetes, hypertension, PCOD, stress, and depression are preferred to get treated by alternative medicine. While insomnia, heart diseases, obesity, allergies, and constipation are preferred to get treated by allopathy medicine.

Objective 1: To assess the socio-economic factors that influence the choice of Alternative medicine and Conventional medicine.

Table 1: Chi-Square test of Socioeconomic factors and choice of type of medicine.

	Value	**df**	**Significance (2-sided)**
Gender	1.036	1	.309
Age	9.780	3	.021
Educational Qualification	2.106[a]	3	.551
Occupation	4.295	3	.231
Income per annum	2.218[a]	3	.528

The above table shows the Chi-Square test of socioeconomic factors and choice of medicine. Gender (0.309) > 0.05, Age (0.021) < 0.05, Educational Qualification (0.551) > 0.05, Occupation (0.231) > 0.05, and Income per annum (0.528) > 0.05.The results show that there is no significant difference between gender, educational qualification, occupation, income per annum, and choice of medicine to treat lifestyle disorders except Age.

Therefore, accepting H_0 and rejecting H_1 for gender, educational qualification, occupation, and income per annum. While for Age we accept H_1 and reject H_0.

The above cluster figure shows the distribution of data into two clusters. Cluster 1 is the distribution of allopathy medicine users, which shows that 86.8 percent of allopathy medicine users are females, among whom 51.5 percent are graduates, 41.2 percent are employed, 73.5 percent are between the ages of 15-30, and 58.8 percent earn below 2.5 lakhs per year. Also showing 70.6 percent of people were cured under the allopathy medicine and the average money spent on consultation is 568.3 and 3,107.35 on medicine.

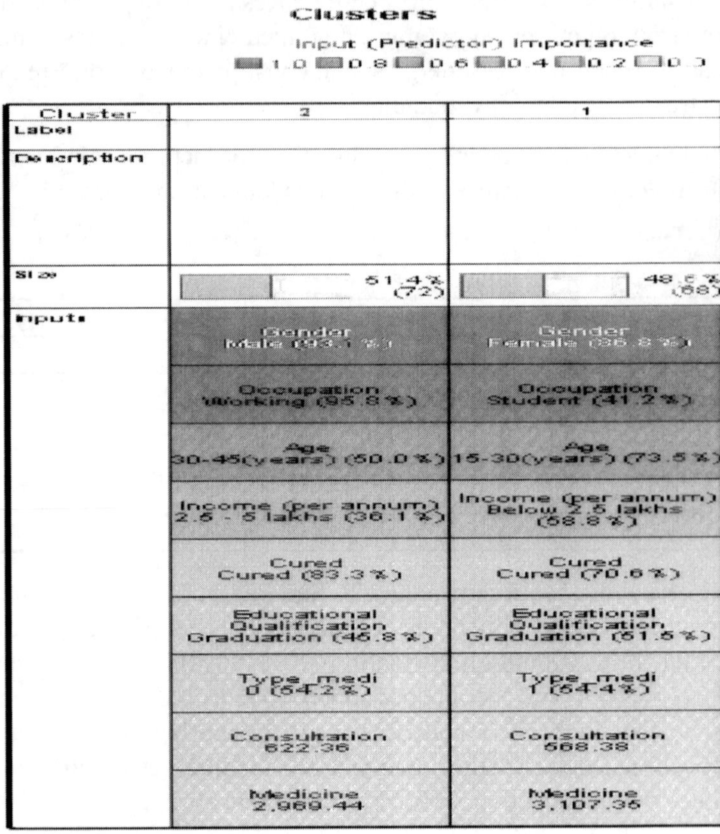

Fig. 7: Cluster figure showing data distribution of Alternative and Allopathy medicine.

Cluster 2 is the distribution of alternative medicine, which shows that 93.1 percent of alternative medicine users are men, 45.8 percent are graduates, 95.8 percent are employed, 50.0 percent are between the ages of 30-45, and 36.1 percent earn between 2.5 and 5 lakhs per year. Also showing 83.3 percent of people were cured under alternative medicine and the average money spent on consultation is 622.3 and 2,969.44 on medicine.

Objective 2- To analyze the cost and effectiveness involved in both practices in treating lifestyle disorders.

Cost-effectiveness analysis (CEA) is a form of economic analysis comparing the prices and results of alternative medical treatments. Clinical outcomes are frequent examples, including blood pressure and cardiovascular events. In this study, we're looking at treating lifestyle disorders such as diabetes, hypertension, heart disease, PCOD, and depression. The respondents were asked about the above disorders following any medical treatment. They were asked whether they had "recovered" or "not recovered" from any illness they had received treatment for.

Table 2: Shows the average cost of allopathy and alternative medicine

	Column Labels	
	0	1
Average Total cost	3490.714286	3774.428571

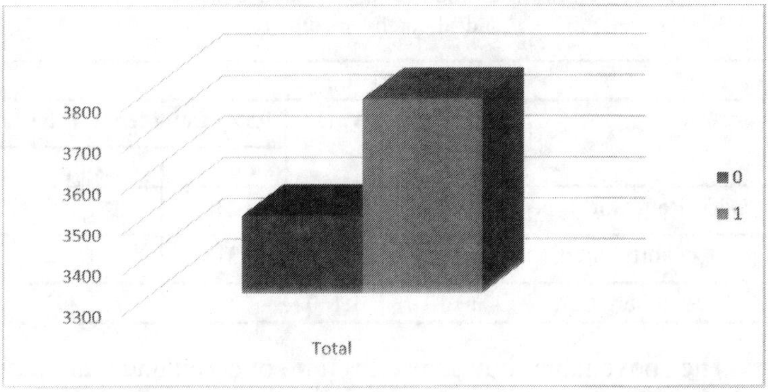

Fig. 8: Average cost of allopathy and alternative medicine.

The above Table 2 and Figure 8 show the average cost of treating lifestyle disorders in allopathy and alternative medicine. Allopathy is assigned 1 and alternative medicine is assigned 0, we can interpret that the average cost of alternative medicine is 3490.71 whereas allopathy has 3774.428 which say that the average cost of treating lifestyle disorder is less in alternative medicine compared to allopathy. We may deduce that the average cost of alternative medicine is 3490.71, whereas allopathy has a cost of 3774.428, implying that alternative medicine has a lower average cost of treating lifestyle disorders than allopathy.

Table 3: Shows the number of people who got cured in treatment of lifestyle disorders under allopathy and alternative medicine

Not Cured		Cured		Total
		Cured		
Type_medi	0	12	58	70
	1	20	50	70
Total		**32**	**108**	**140**

The above figure shows the number of people who got cured under alternative and allopathy medicine. Here it is shown that the number of cured under alternative is greater than the number of cured under allopathy.

Table 4: Shows the odd ratio for cured under alternative and allopathy medicine.

Risk Estimate			
	Value	**95% Confidence Interval**	
		Lower	**Upper**
Odds Ratio for Type_medi (0/1)	.517	.230	1.162
For cohort Cured = Not Cured	.600	.318	1.132
For cohort Cured = Cured	1.160	.966	1.392

The above table shows the odds ratio of cured under allopathy and alternative medicine, the ratio of 1.160 shows that there is a 16 percent greater chance of getting lifestyle disorders cured under alternative medicine than in allopathy.

CER could be determined only for a few disorders, according to the proposed definition of health effect, because in most cases this ratio is inestimable due to zero recovery rates. However, instead of "full recovery", the "degree of recovery" is considered. The above table shows the cost and effectiveness ratio for the treatment of lifestyle disorders under alternative and allopathy medicine. The ratio shows that the alternative is slightly better off in cost and effectiveness than allopathy medicine.

Table 5: Shows the cost-effective ratio (effectiveness is the recovery rate)

Cost and effectiveness ratio of lifestyle diseases treatment for treatment under	
Allopathy medicine	Alternative medicine
4887	4555.34

Table 6: ANOVA test for cost and effectiveness in alternative and allopathy medicine

Tests of Between-Subjects Effects					
Dependent Variable: Total_cost					
Source	**Type III Sum of Squares**	**df**	**Mean Square**	**F**	**Sig.**
Intercept	1249790305.405	1	1249790305.405	351.984	<.001
Type_medi	32648.060	1	32648.060	.009	.924
Cured	24094.085	1	24094.085	.007	.934
Type_medi * Cured	4975994.691	1	4975994.691	1.401	.239
a. R Squared = .017 (Adjusted R Squared = -.005)					

The above Table 6 shows the p-value (0.239) interpreting that there is no significant difference in cost and effectiveness between alternative and allopathy medicine in treating lifestyle disorders. The R Square of 0.17 states that 17 percent of the data is explained and 83 percent of the data is unexplained.

Hence, accepting H_0 and rejecting H_1.

Objective 3- To assess the funds released by the government towards alternative medicines and allopathy.

The Table 7 and Figure 9 shows the budget allocation towards allopathy and alternative medicine. The trend line was run to see the trend of budget allocation, the upward-sloping trend line shows that there is an increase in allocation each year towards both forms of medicine. Even though the allocation towards alternative medicine started in the eighth plan, there is still an increase in allocation towards the growth of alternative medicine.

Table 7: Fund allocation of the Twelve plans to allopathy and alternative medicine

Year	Health & Family welfare	AYUSH
1951-1956	65.3	
1956-1961	145.8	
1961-1966	250.8	
1966-1969	210.6	
1969-1974	613.5	
1974-1979	1252.6	
1979-1980	341.6	
1980-1985	3412.2	
1985-1990	6809.4	
1992-1997	13994.2	108
1997-2002	34938.6	266.35
2002-2007	58145.3	775
2007-2012	136147	3988
2012-2017	75145.3	10044

*In crores

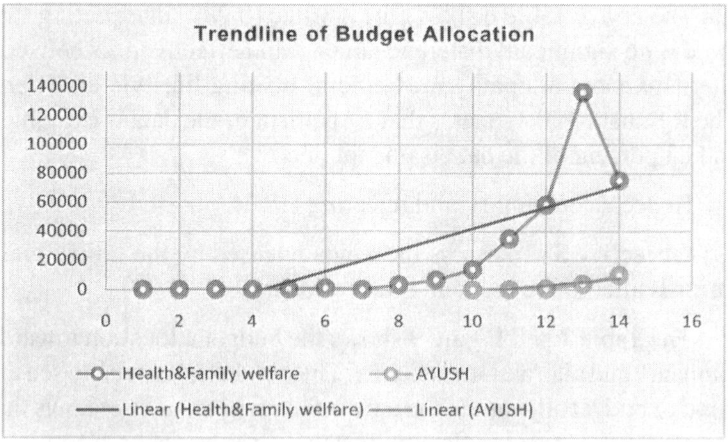

Fig. 9: Trendline of budget allocation.

Table 8: Annual allocation of funds towards allopathy and alternative medicine

Year	Health & Family welfare	AYUSH
1991-1992	1898.8	
2012-2013	6585	990
2013-2014	8166	1069
2014-2015	8233	1069
2015-2016	11324.5	1008
2016-2017	14824.6	1050
2017-2018	17661.8	1428.65
2018-2019	20670.4	1626.37
2019-2020	23108.12	1939.76
2020-2021	25172.8	2122.08
2021-2022	28291.27	2970.3

*In crores

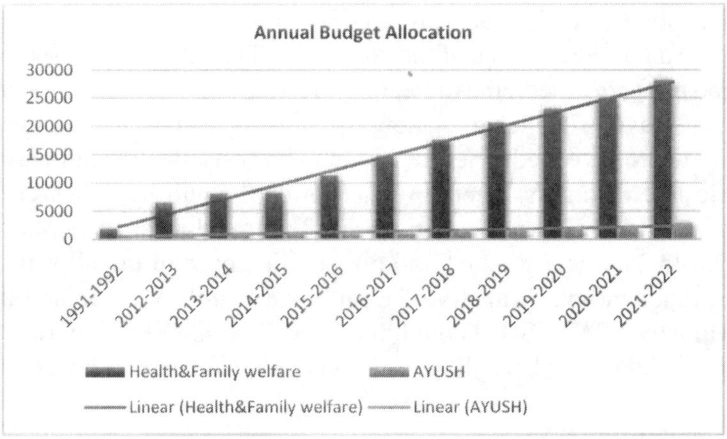

Fig. 10: Trendline of annual budget allocation.

The Table 8 and Figure 10 shows the annual budget allocation for allopathy and alternative medicine. The annual plans have a positive upward trend for allopathy while for alternative medicine the trendline is a slightly upward-sloping line, showing slightly less allocation compared to allopathy medicine.

Discussion

The primary data shows that Ayurveda is preferred more followed by Homeopathy, Yoga, and Unani in treating lifestyle disorders. It showed that diabetes, hypertension, PCOD, stress, and depression are preferred to get treated by alternative medicine. While insomnia, heart diseases, obesity, allergies, and constipation are preferred to get treated by allopathy medicine. It depicted that people's preferences for alternative medicine are mainly due to no side effects, natural/organic, and efficient. Allopathy medicine is preferred by people mainly as they are convenient, available, and gives fast results. Chi-Square was run in the analysis of the socio-economic factors that influence the choice of Alternative medicine versus Conventional medicine, no significant differences in gender, educational qualification, occupation, or income per capita were discovered in the choice of medicine, except age, which revealed a significant difference in the choice of Alternative medicine versus Conventional medicine. Therefore, accepting H_0 and rejecting H_1, except age which is accepting H_1 and rejecting H_0. ANOVA test was run to the analysis of the cost and effectiveness of alternative and allopathy medicine in treating lifestyle disorders showing that there was no significant relationship, showing that neither allopathy nor alternative medicine was found to be cost-effective in treating lifestyle disorders. Therefore, accepting the null hypothesis and rejecting the alternative hypothesis. The trendline analysis shows that there was a gradual and positive increase in the allocation of funds in both Ministry of Health and Family Welfare and the Ministry of AYUSH. Though the allocation in AYUSH started in the eighth-year plan still has seen a positive increase in funds.

Conclusion

Lifestyle disorder has been the most important concern to the population with the pandemic there has been an increase. keeping SDG's goal of ensuring health and well-being for all people of all ages, this paper shows the effect of socioeconomic and cost-effectiveness ratios in treating lifestyle disorders under allopathy and alternative medicine. This result shows that there was a significant difference seen in the age in the choice of medicine in treating their ailment compared to gender, income, qualification, and occupation, which shows insignificance. This study also shows that the cost-effectiveness ratio was slightly seen better in alternative medicine than in allopathy medicine. Merely putting in more attempts to curb costs so that everyone, regardless of age, will have access to receive care, achieving the sustainable goal.

References

Das, T. K. (2005). A Field Survey to Evaluate Costs and Effectiveness of Complementary and Alternative Medicines. *DSA Centre for Regional Economics,* Jadavpur University, Kolkata.

Naikwade, S., Gopal, R., & Sippy, N. (2016). A Study on Working Women's Attitude Towards Allopathy or Ayurveda & Yoga System of Medicine For Managing the Lifestyle Diseases With Special Reference To Western Suburbs Of Mumbai. *International Journal of Business Management & Research (IJBMR)*, *6*(1), 57-64.

Egede, L. E., Ye, X., Zheng, D., & Silverstein, M. D. (2002). The prevalence and pattern of complementary and alternative medicine use in individuals with diabetes. *Diabetes care*, *25*(2), 324-329.

Tavares, A. I. (2015). Substitutes or complements? Diagnosis and treatment with non-conventional and conventional medicine. *International Journal of Health Policy and Management*, *4*(4), 235.

https://www.ncbi.nlm.nih.gov/pmc/articles/PMC4020364/

https://www.britannica.com/science/complementary-and-alternative-medicine

https://guides.library.duq.edu/complementary_medicine/history

https://www.verywellhealth.com/what-is-allopathic-medicine-2249039

https://diabetesatlas.org/

https://www.who.int/news/item/04-03-2022-world-obesity-day-2022-accelerating-action-to-stop-obesity

https://www.who.int/india/health-topics/hypertension

https://www.who.int/india/health-topics/hypertension

https://www.statista.com/statistics/1136572/india-polycystic-ovary-syndrome-issues-among-women-by-age-group/

https://www.who.int/india/health-topics/depression

https://www.cbhidghs.nic.in/index1.php?lang=1&level=1&sublinkid=75&lid=1135

https://main.mohfw.gov.in/

https://www.ayush.gov.in/

7

Impact of Covid-19 on the Health of the Chenchus of Nallamala Hills of Andhra Pradesh

Sagar Kodi, Kondaiah Nayak Malavathu and *Valerie Dkhar*

Introduction

The Corona Virus (Covid-19), caused by the SARS COV-2, from Wuhan, China (Forster et al. 2020), is recognised as a significant threat to public health and on 11 March 2020, the WHO declared it a global pandemic (Menon & Schmidt-Vogt, 2022). Its drastic effect was felt globally and to a greater degree in densely populated nations that do not have adequate livelihoods and healthcare facilities (Masrur et al. 2020) which also holds in the Indian scenario. The situation was significantly challenging to tackle due to inadequate health infrastructure in rural areas and insufficient trained medical personnel in India. Around 60-65 percent of the population lives in rural areas with fewer healthcare facilities than in cities (Mitra, 2020). The government decided to impose several spells of lockdowns and ban travel and everyday activities to contain the spread of the virus. This was done to prioritise people's health over their livelihood (Kumar et al. 2020) after the first Covid-19 case in Kerala on 27 January 2020 was detected (Aggarwal et al. 2012). Even after taking precautionary steps like lockdown and travel ban in the country, increasing Covid-19 infections created a new challenge due to the scarcity of testing services, weak surveillance, and poor medical care (Kumar et al. 2020). The Chief Scientist of WHO, Dr. Soumya Swaminathan, has expressed concern and warned that rural India might become the next coronavirus hotbed and emphasised the need to use this opportunity to strengthen and improve its primary healthcare system (E Kumar Sharma, 2020).

Imposing lockdown in the country affected people from all walks of life, some worse than others. It delayed the treatment for non-Covid-19 patients making it difficult to access medical care. Most healthcare resources shifted to Covid-19 management, and they feared falling ill during Covid-19 times during check-ups (Hebbar, et al. 2020). The Covid-19 pandemic spread beyond borders and affected not only the physical health of the people but also their mental health. All medical and technical staff were diverted to fight the virus, leaving little to cater to other medical cases and emergencies that brought about a lot of stress to the people. Before Covid-19, Sustainable Development Goals initiated progress in improving health and fighting against Non-Communicable Diseases (NCDs) (Alisjahbana, 2020). However, during the pandemic, disruptions and suspension of health-related campaigns like polio, measles, HIV, and RMNCH+ services meant for pregnant women and children have reversed the progress achieved till now in healthcare services, mainly in maternal and child healthcare services in reaching the targeted SGDs (Bhatia and Khetrapal, 2020).

The Chenchus, listed as a Particularly Vulnerable Tribal Group (PVTG), inhabit the interior forests of the Nallamala hills of Andhra Pradesh, have also faced innumerable difficulties as a consequence of Covid-19, viz., health issues, loss of livelihood, etc. This paper aims to critically analyse the impact of Covid-19 on the health status of the Chenchus of Nallamala Hills based on primary data gathered from the field, which is aptly substantiated by secondary sources. It also intends to give a comprehensive idea of the health status of the Chenchus as well as investigate the post-Covid situation among them.

Method: Interviews were conducted from February to May 2022, and data about healthcare facilities and details about the Chenchu tribe were collected from July to October 2022. Informed consent was taken from individuals, and permission was taken from Primary Healthcare Centres (PHCs) in the Atmakur division, Kurnool district of Andhra Pradesh (AP).

Interview Schedule: An interview schedule was developed. Participants were asked to tell about their personal experiences regarding livelihood, the health impact of Covid-19, and associated challenges. The schedule checks for comprehension of impact, accessibility to medical assistance, and data consistency. The researchers collected all the items listed in the schedule to calculate data accuracy.

Data Collection: Data was collected directly from Chenchus and Primary Healthcare Centres (PHCs) in the Atmakur Division, Kurnool District of Andhra Pradesh. Personal interviews and case studies were taken using the interview schedule. The researchers also reached out to each PHC and collected secondary data related to Covid-19 and healthcare facilities utilized by the Chenchus. The sample size of 102 interviews was conducted on the Chenchu households falling in the Atmakur Division of Kurnool District of Andhra Pradesh.

Data Analysis:MS Excel and a Percentage calculator was used to analyse the secondary data about Covid-19 positives, institutional deliveries, immunization, and healthcare facilities used by the Chenchus.

Results: A total of 102 families, consisting of more than 300 members with varying experiences regarding the impact of Covid-19 on their livelihood, and health, both physical and mental, which was stress-induced, were interviewed.

Table 1: Total Population of the Villages

S No	Village Name	Male	Female	Total	Sub Centre	PHC	Mandal
1	Bairluty	207	277	484	Bairluty	Bairluty	Atmakur
2	Naguluty	127	147	274	Bairluty	Bairluty	Atmakur
3	Palem-cheruvu	115	114	229	Guvvalakunta	Kothapalli	Kothapalli
4	Chadha-rampenta	69	58	127	Guvvalakunta	Kothapalli	Kothapalli
5	Janalagudem	78	68	146	Yerramatam	Yerramatam	Kothapalli
6	Yerramatam	120	155	275	Yerramatam	Yerramatam	Kothapalli
7	Total	716	819	1535			

Source: Fieldwork, 2022.

Table 2: Villages where Data was collected

S. No	Primary Healthcare Centre	Sub-Centres	Villages Come Under
1	Bairluty	Bairluty	Bairluty and Naguluty
2	Yerramatam	Yerramatam	Yerramatam, Janalagudem, and Pathamadugulagudem.
3	Kothapalli	Guvvalakunta	Palemcheruvu, Chadharampenta

Source: Fieldwork, 2022.

Covid-19 Cases and Deaths

Some of the Chenchus were affected by Covid-19. However, they were asymptomatic. Unlike the rest of the people, Chenchus were not severely affected by it. Since they mainly lived isolated from the rest of the population, they were less impacted. They are more independent, do not mingle with others, and prefer not to go outside their community. At the same time, they also do not allow outsiders into their communities and villages. As a preventative measure, if outsiders choose to come for various purposes, they have not allowed them inside their villages to prevent the spreading of the virus.

From the data collected from 7 villages, only 38 tested positive, with zero deaths. It shows that their measures to isolate themselves from outsiders were highly impactful in avoiding any deaths among them.

Table 3: Covid-19 Cases and Deaths among the Chenchus

Sl. No	Village Name	Covid-19 Positives	Deaths
1	Bairluty	1	0
2	Naguluty	1	0
3	Chadharampenta	15	0
4	Janalagudem	2	0
5	Yerramattam	3	0
7	Palemcheruvu	13	0
8	Total	38	0

Source: Fieldwork, 2022.

Institutional Deliveries among Chenchu of Biarluty and Naguluty: Data on institutional deliveries before COVID-19 in 2019 (31 deliveries) and during the peak of the pandemic in 2020 (41 deliveries) were compared. We observe an increase in institutional deliveries which points to the fact that COVID-19 has had no impact on institutional deliveries. This can be attributed to the continuous and dedicated efforts of the ASHA workers who have provided counsel and care to pregnant women pre and post-delivery.

Table 4: Institutional Deliveries among Chenchus

S. No.	Village	Before Covid-19 (2019)	During Covid-19 (2020)
1	Bairluty & Naguluty	14	17
2	Yerramatam gudem	7	9
3	Palemcheruvu	6	10
4	Chadharampenta	1	4
5	Janalagudem	3	1
6	Total	31	41

Source: Fieldwork, 2022.

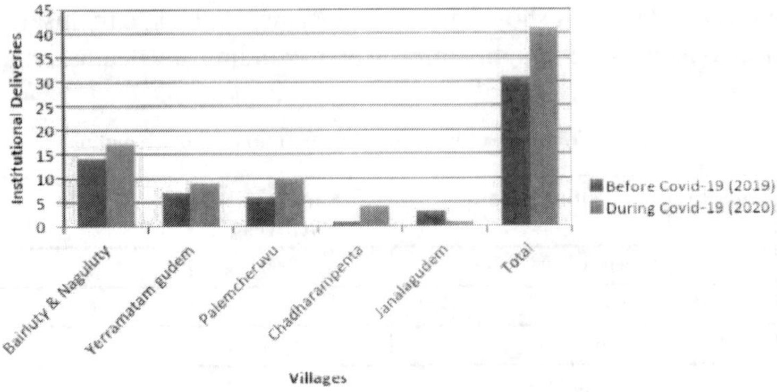

Fig 1: Institutional Deliveries among Chenchus.

Immunization for the Children: After childbirth, the immunization rate increased to 20 percent in 2020 compared to 2019 among Chenchu. Since they feared Covid-19, they became more cautious about other possible viruses and vaccinated their children. PHCs also created more awareness about the importance of getting vaccinated for other health concerns, motivating them to vaccinate their children promptly.

Table 6: Immunization for the Children:

S. No	Village Name	Before Covid-19 (Oct-Dec 2019)	During Covid-19 (Apr-June 2020)
1	Bairluty	9	12
2	Palemcheruvu	4	3
3	Chadharampenta	2	2
4	Naguluty	5	7
5	Total	20	24

Source: Fieldwork, 2022.

Challenges: Approaches to Healthcare

The risk factor of Covid-19, along with the unknown factor of its rapid spreading to many people, caused the Chenchus to feel concerned. This has resulted in a decline in visiting healthcare facilities than before. Rural Medical Practitioners (RMPs) generally come to villages to treat the sick. Hence, they prefer to visit RMPs instead of going to government hospitals.

Before Covid-19, for regular health check-ups, Chenchus would go to the RMP hospitals nearby the town of Atmakur. However, during Covid-19, for fear of contracting the virus and the consequences after, they prefer not to. This has resulted in frequent visits to the PHCs nearby their villages rather than to the RMP hospitals. Some of them never tried to visit hospitals during the pandemic. During Lockdown, PHCs conducted awareness programmes, government-sponsored Covid tests, and carried Covid-19 vaccination drives in the Chenchu villages. They also provided masks, sanitisers, and soaps. Despite these facilities, when comparing the 2019 and 2020 statistics, male health-related check-ups in PHCs decreased by 40.25 percent. For women -32.92 percent decreased during 2020 compared to 2019. Overall, for the year 2020, it declined to -36.76 percent.

Table 5: Health-Related Check-ups in the PHCs

S No.	Village Name	Before Covid-19 (Oct-Dec 2019)		During Covid-19 (Apr - June 2020)	
		Male	Female	Male	Female
1	Biarluty	1082	968	569	604
2	Naguluty	186	174	122	105
3	Palemcheruvu	91	68	102	96
4	Chadharampenta	29	32	32	25
5	Yerramatam	25	29	19	21
6	Janalagudem	23	32	14	23
7	Total	1436	1303	858	874

Source: Fieldwork, 2022.

Challenges: Loss of Livelihood and its impact on Chenchu's Health

Covid-19 indirectly impacted the Chenchus in the Nallamala regions of Kurnool District, Andhra Pradesh. During the lockdown, they lost their primary source of income via, i.e., Mahatma Gandhi

National Rural Employment Guarantee Act, 2005 (MGNREGA) because this work was designed to be completed as a group. There was a legitimate health concern that if a group member tested positive for Covid-19, others might get infected. Given the consequences, this programme was paused temporarily, so with this loss of the MGNREGA livelihood opportunity, their stable financial support was taken away from them, which resulted in financial problems in meeting their daily needs.

Apart from not having enough money to buy food, groceries, and essentials, they could not afford to visit private hospitals for health check-ups. Their other source of income was the sale of Minor Forest Products (MFP). However, during the lockdown, sales of the MFPs reduced consequently due to the fear of establishing contacts.

Women from the Chenchu community also engage as daily labourers in the agricultural fields. However, during the lockdown, they lost their work because of suspended cultivation and mainly depended on pensions and financial support from the Government of Andhra Pradesh.

Challenges: Food and Chenchu's Support System

During the lockdown, several non-government organizations (NGOs) like the Evangelical Fellowship of India Commission on Relief (EFICOR), Koval Foundation, and Rural Development Trust (RDT) came forward to help them with items like rice bags, groceries, masks, gloves, and sanitisers. They were also provided ration and a few other essential groceries from the state government, which sustained them during the peak of covid lockdown. In some cases, however, larger families had to survive by eating rice with pickles, chilli powder, and Ganji (rice gruel) as groceries finished sooner than anticipated. Despite experiencing lifestyle changes, they were able to come out of it resilient.

Conclusion

The pandemic has brought new rules we must adhere to survive and sustain ourselves. Lives have been lost, families separated, and livelihoods disrupted. The Chenchus find themselves in a very precarious position as they have been able to isolate themselves to prevent contracting the virus, yet, they were not left untouched by it. Even though very few positive cases and no deaths were recorded, they faced an uncertain future. The suspension of the MGNREGA scheme and the decline in the sale of MFPs caused due to curtailed movements and restrictions imposed during the pandemic and subsequent lockdowns led to disruptions in their economic life. However, due to their location and timely intervention by the government and non-government organizations (NGOs), the infection and death rates among them were lower than the national average. Today they are slowly getting back on their feet, and it will take concerted efforts from all fronts, the people, the various NGOs working in their area and the state government to ensure the sustainability of their way of life.

References

Andrews, M. A., Areekal, B., Rajesh, K. R., Krishnan, J., Suryakala, R., Krishnan, B., Murray, C. P., & Santhosh, P. V. (2020). First confirmed case of Covid-19 infection in India: A case report. *Indian Journal of Medical Research*, 151(5), 490. https://doi.org/10.4103/ijmr.ijmr_2131_20

Aggarwal, A., Mehta, S., Gupta, D., Sheikh, S., Pallagatti, S., Singh, R., & Singla, I. (2012). Clinical & immunological erythematosus patients characteristics in systemic lupus Maryam. *Journal of Dental Education*, 76(11), 1532–1539. https://doi.org/10.4103/ijmr.IJMR

Ali, M, Bhuiyan, M. (2020). Around 1.7m youths may lose jobs in Bangladesh in 2020 for a pandemic. (n.d.). Retrieved 23 November 2022, from https://www.reddit.com/r/autotldr/comments/ickslI/around_17m_youths_may_lose_jobs_in_bangladesh_in/

Alisjahbana, A. S. (2020). Asia and the Pacific SDG progress report. In United Nations Publication. https://www.unescap.org/publications/asia-and-pacific-sdg-progress-report-2020

Bhatia, R., & Khetrapal, S. (2020). Impact of covid-19 pandemic on Health System & Sustainable Development goal 3. *Indian Journal of Medical Research*, 151(5), 395. https://doi.org/10.4103/ijmr.ijmr_1920_20

Creswell, J. W., & Poth, C. N. (2018). Qualitative inquiry and research design: Choosing among five approaches. Sage.

Dhingra, S. and Ghatak, M., (2022, May 12). How has Covid-19 affected India's economy? Economics Observatory. Retrieved 23 November 2022, from https://www.economicsobservatory.com/how-has-covid-19-affected-indias-economy

E Kumar Sharma. (2020). Rural India's next coronavirus hotbed, warns WHO chief scientist. (n.d.). Retrieved 23 November 2022, from https://www.businesstoday.in/opinion/interviews/story/rural-india-next-coronavirus-hotbed-warns-who-chief-scientist-253300-2020-03-29

Forster, P., Forster, L., Renfrew, C., & Forster, M. (2020). Phylogenetic network analysis of SARS-CoV-2 genomes. *Proceedings of the National Academy of Sciences of the United States of America,* 117(17), 9241–9243. https://doi.org/10.1073/pnas.2004999117

Hebbar, P., Sudha, A., Dsouza, V., Child, L., & Amin, A. (2020). Healthcare delivery in India amid the Covid-19 pandemic: Challenges and opportunities. *Indian Journal of Medical Ethics*, 05(03), 215–218. https://doi.org/10.20529/ijme.2020.064

Kumar, A., Nayar, K. R., & Koya, F. (2020). Public Health in Practice Covid-19 : Challenges and its consequences for rural healthcare in India. 1(May), 1–2. https://doi.org/10.1016/j.puhip.2020.100009

Masrur, A., Yu, M., Luo, W., & Dewan, A. (2020). Space-time patterns, change, and propagation of Covid-19 risk relative to the intervention scenarios in Bangladesh. *International Journal of Environmental Research and Public Health*, 17(16), 1–22. https://doi.org/10.3390/ijerph17165911

Menon, A., & Schmidt-Vogt, D. (2022). Effects of the Covid-19 Pandemic on Farmers and Their Responses: A Study of Three Farming Systems in Kerala, South India. *Land*, 11(1). https://doi.org/10.3390/land11010144

Rawat, D., Dixit, V., Gulati, S., Gulati, S., & Gulati, A. (2021). Diabetes & Metabolic Syndrome : Clinical Research & Reviews Impact of Covid-19 outbreak on lifestyle behaviour : A review of studies published in India. *Diabetes &Metabolic Syndrome: Clinical Research & Reviews,* 15(1), 331–336. https://doi.org/10.1016/j.dsx.2020.12.038

Sharma, G.D. & Mahendru, M. 2020. Lives or livelihood: Insights from locked-down India due to Covid-19. *Social Sciences & Humanities Open* 2, 100036. https://doi.org/10.1016/j.ssaho.2020.100036

Vora, R. (2021, 5 August). The implications of Covid-19 for Rural India. India Development Review. Retrieved 23 November 2022, from https://idronline.org/the-implications-of-covid-19-for-rural-india/

8

Effectiveness of Online Teaching and Learning Within the College during the Lockdown–A Study

M. Deepa

Introduction

Covid-19 has thrown the entire world into a quagmire, with no apparent solution in sight. Wuhan, China, was the site of the first recorded outbreak. This infectious disease was officially named Covid-19 and SARS-CoV2 on February 11, 2020. The original cause of viral transmission to humans, as well as whether the virus became pathogenic before or after the spillover case, are still unknown. When a person who is infected coughs, sneezes, or talks loudly, hundreds of millions of virus particles are released into the environment through respiratory droplets, where they can remain infectious for days. Smaller droplets will stay suspended in the air for up to 20 minutes or up to 48 hours on surfaces indoors, while larger droplets settle on all surrounding surfaces Covid-19 may be caught by others. The Covid-19 pandemic has a thought-provoking impact on the Sustainable Development Goals (SDGs) that were approved by the United Nations in 2015 (Cf, 2015). Out of these goals, the fourth goal of the SDGs has been formulated to address the global level of education, in terms of 'Quality Education'. This particular goal propagates that every learner must obtain the skills and knowledge that is required to endorse sustainable development in the world, for the betterment of the global population, the future generations, and the environment. However, the advent of Covid-19 has affected the education sector of the entire world, primarily due to the closure of educational institutions. The Covid-19 pandemic still remains a serious challenge for the entire humanity, and unfortunately, the world is still in its firm grip. Throughout the world as we know it, every aspect of life has completely been altered with the spread of this virus. In such times, the prime aim of all the countries is to decrease the rapid spread of Covid-19 in the society. Therefore, as a drastic measure to curtail mass infection, schools, colleges, universities, and other higher education institutions worldwide were closed due to the spread of the deadly Covid-19. As a result, policy-makers adopted to the new normal as rapidly as they could, especially in the global educational sector. As of the current scenario as well, most of the educational institutions worldwide have adapted to the online classes, so as to continue the process of teaching and learning which can have a great benefit to achieve 2030 SDGs.

The whole nation came under this coronavirus, people started questioning the government and, in the meanwhile, the education system was also abruptic. The Tamil Nadu government on 23 March 2020 announced a lockdown in the state. Apart from essential services, other things were announced to be closed. As the Covid-19 pandemic spread, schools, colleges and, universities were forced to shut down and there has been an increasing move towards teaching online for an indefinite time. Therefore, it was the time to rethink, reframe and rebuilt the education system and fulfil the requirement of not only the learners but also to provide the best platform for the next generations. In the mid of this ruckus, college teachers have to decide which platform they think is appropriate for the learners. Moreover, learners have to be ready to adapt to this changing technology and the whole paradigm of education. Justice Basheer Ahmed Sayeed College for Women, situated in Teynampet, Chennai is also known as JBAS College for Women (Autonomous).The aim of establishing this college was to impart higher education to social and economically weaker students of the city. The college was established in 1955 and has been recognised by the University Grants Commission affiliated with the University of Madras and accredited by NAAC. At the JBAS College, both teachers and learners quick adapt to this transition from offline to online mode.

The teachers started exploring all forms of platforms available online. Platforms like Zoom, WebEx, and Google meet, and Google classroom was used initially. Later an online portal through which the teachers keep a record of personal and student progress information was improvised to communicate with the students. For this same reason, the Computer department of the college developed and launched a new platform LMS (Learning Management System) for catering to teachers and learners. This Research highlights the teacher's and learners' perceptions of the effectiveness of these tools and the problems faced by students and teachers during a lockdown. Despite all the problems would they recommend having hybrid teaching-learning post-pandemic?

Objectives

- To explore the various modes of online teaching and learning during the lockdown period.
- To examine the teacher's and learner's knowledge of digital tools used during the lockdown period.
- To study the views and problems of teachers and learners faced in the online learning system during the lockdown.

Literature Review

The data used in this study were mainly drawn from research and literature on online teaching and learning during the Covid-19 pandemic. A large portion of the planet, in particular, is under quarantine due to a significant outbreak of the Coronavirus. It's an infectious disease that spreads when you come into contact with someone who has already contracted it. As a result, the disease-affected areas have been put under lockdown. Education was changed to an online mode as a result. Online learning is characterised as learning through the use of various devices (e.g., cell phones, laptops, etc.) with adequate internet access, in which students are free to communicate with teachers and peers. Over the past few decades, a significant number of alternative methods have been created to deal with issues like this if they ever arise. In developing countries, e-learning implementation faces a large number of challenges, primarily due to teachers' and institutions' limited resources, both in terms of time and money. Online learning is an interactive and integrated collection of tools that provides educators, learners, and others involved in education with knowledge, tools, and resources to support and enhance educational deliberations, according to SAP Litmos, an award-winning LMS. At the same time, E-learning does not have to be an all-or-nothing proposition, while in a conventional classroom, teachers must devote a significant amount of time to preparing classes, and students must conduct extensive study prior to submitting assignments. Gmail, Google Meet, Google Classroom, and Zoom Classroom are just a few of the online options. These platforms can also be used to simplify the teaching-learning process. Teachers and students took the help of podcasts, blogs, and vlogs. Students also use Power Point presentations, books, Internet sources, marketing dictionaries, and assignments. All student activities are documented, and students receive grades for everything they do. Grades are assigned based on their activity and level of engagement. In the classroom, only midterm exams are given. There are several ways to improve grades.

Methodology

The researcher used both qualitative and quantitative methodologies to investigate teachers' and students' attitudes toward online tools and to identify the reasons for high exam scores. The scope of this research is limited to the History Department of Justice Basheer Ahmed Sayeed College for Women.

Participants

This study included all the students of the History Department of JBAS College. The study included approximately 100 students. Teachers were asked to share their online teaching experiences, and students were asked to share their reasons for achieving high grades in online mode. Aside from that, problems encountered by students and teachers are addressed, and possible solutions are proposed.

Design

The data for this study was gathered using a questionnaire designed for learners to investigate their perceptions of the online teaching-learning mode. The questionnaire consisted of ten questions, both qualitative and quantitative, in order to learn about the experiences, perceptions, and reflections of online teaching-learning. All of the respondents gave their complete cooperation in answering all of the questions honestly. Teachers and students were given ample space to freely share their thoughts and experiences for the purpose of research. A semi-structured interview was conducted with teachers to gather the necessary data.

Findings

This section presents findings found from the collected data. The researcher performed a survey study in which percentage analysis was used to investigate the various types of modes used by college teachers and students during the lockdown era in order to obtain the results. For teaching and assessment purposes, teachers and students were given their own college-developed LMS. Several training sessions were held to make the process easier and to assist the teachers and students. The goal was to have a smooth transition between classes. Despite the college's efforts, a significant number of teachers and students have shown an interest in learning through other online options. Initially, teachers used WhatsApp for communication before realising that this was not a temporary issue. Teachers then used various online platforms to perform the lessons. The Google Classroom platform is well-liked by both teachers and students. It was more dependable than the LMS and Zoom platforms, according to them. Many teachers used Google Classroom for online classes, class assessments and assignments.

However, for lectures, most of the teachers used Google Classroom, citing that it was a simple platform to use and that it enabled them to keep track of student attendance. Afew teachers said they used the Zoom platform for classes but then switched to other platforms due to safety concerns. Students attended lectures and webinars that were held online. Students were often encouraged to use research materials for additional information by requiring access to e-books and e-journals. MOOC lectures are now a standard part of a college education. Many other digital platforms have been launched to help teachers and students save time.

Findings relevant to the teachers' and students' knowledge of online resources. A semi-structured interview found that both teachers and students were wary of using online resources. The evolving pedagogy model did not sit well with teachers and students. Many of the students, who came from far-flung parts of the city, struggled to get to class on time, and they often skipped the entire lecture. Students' main factors were that they lacked basic knowledge of how to use online platforms and that they did not have compatible networks. But it wasn't just students who were resistant to online learning; many faculty members were as well. Maybe they haven't used this forum before or they don't want to alter it. Despite majority of teachers appear to be computer-literate enough to teach online classes. Online lectures were found to be more successful by many of the teachers than live classroom lectures, and the same teachers were pleased with the student-teacher interaction during online teaching and learning. They said that changing one's environment is as natural as changing one's clothing. Some teachers claimed that they find the online mode to be quite comfortable as they have already been doing online courses along with their students through MOOCs and NPTEL lectures. One more point highlighted by the teachers and students is that now-a-day's almost everyone has advanced machinery, so accessing classes was not at all an issue unless they have a high-speed internet connection.

Findings of the teachers' and students' perspectives on the overall online teaching-learning experience and the problems faced them. These results shed light on the experiences of teachers and students during the Covid-19 pandemic. Teachers although quick and adaptive to these changes complained that they didn't find the home environment comfortable in preparing and delivering lectures. Teachers were unhappy with the lockdown as all the schools and colleges were closed at the drop of the hat, which didn't give female teachers enough time to adjust to their small children and at the same time they were required to fulfil their lectures online. All kind of setup needs time but this situation retreated everyone from their routine work. For teaching purposes, teachers needed secluded places and high connectivity. Once you get settled with these primary conditions, teachers were also worried about their students' presence in the classes which was minimal. Many teachers later come up with the idea of switching from one online platform to another for better interaction with students. Some students from remote areas were still finding it difficult to join the classes. It is found that the college's initiative for online teaching-learning and providing their own software was a relief for the teachers. Many senior teachers who were not tech-savvy, recorded their lectures and shared them with the students but then the files were so heavy that students find it difficult to download them. Many teachers were not fine with the idea of sitting in front of a camera. They think the face-to-face classroom is something they are accustomed to but delivering lectures with a camera on was a different experience. In online classes syllabus was again difficult to complete with the number of teaching hours. So teachers have to curtail the unnecessary or repetitive parts of the syllabus. Sometimes because of internet issues, the voice lagged and the quality time was wasted on fixing the issues. As far as students' opinion is considered, they were of the opinion that online classes were although boring but safe for them at the time of this pandemic where you can protect yourself and your family only by maintaining social distancing. The study also revealed that students were unhappy with online classes for many reasons. The majority of the students felt that they found it difficult to communicate with their teachers and were asked to be quiet in the classroom. If they had any queries, then they have to wait for the next class. Students were asked to share their general feeling while attending online classes, then 56 percent of students found online classes to be difficult to understand and they took the help of other recourses. 22 percent of the students found online classes interesting while 16 percent of the students felt sleepy during classes. Overall, 60 percent of the students complained that they faced trouble in online classes. Students were asked whether the online classes helped them in scoring good marks and 30 percent said no but 70 percent of the students accepted that yes objective tests were easy to perform and they found it scoring. When asked that do they think technology can help in reducing the opportunities for cheating, 48 percent of the students agreed and 20 percent of the students disagreed and 32 percent of students remained neutral.

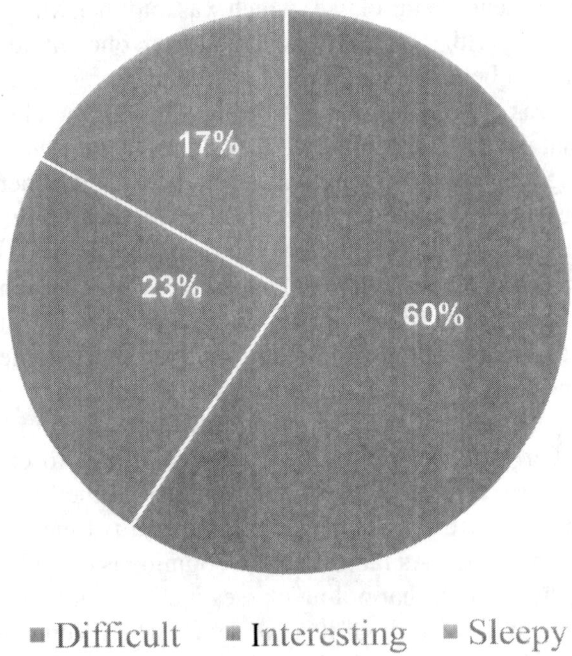

Fig. 1: Online Classes—Student's Understanding

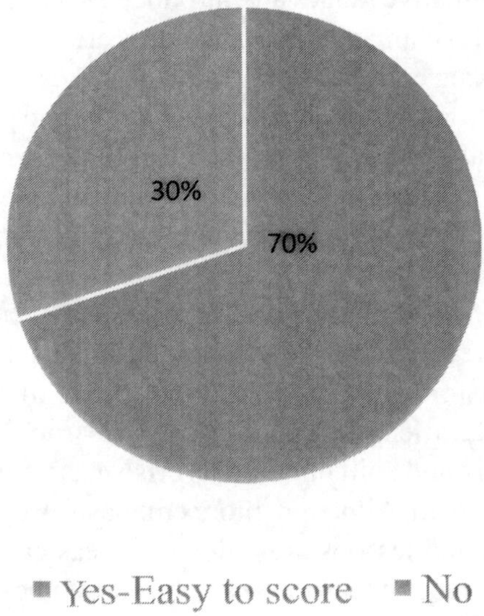

Fig. 2: Online Exams—Better in Scoring marks.

Conclusion

Due to Covid-19, this immediate transition from traditional to online pedagogy education has provided enough time for the nation, in particular, the college to not only recognise the strengths and weaknesses but also, to accept the challenges and turn them into opportunities to be ready in the future. Many colleges were initially resistant to the changes. They were not ready to accept the seriousness of the crisis and no one was expecting this crisis to be long-lasting. Neither teachers nor students were ready for the crisis. Everyone was facing difficulty in changing and breaking the monotonous of this life. Students with bad connectivity were not able to be present in all the classes which resulted in less percentage of attendance. According to the rule of the university, each student has to score 75 percent of attendance. Later when the government had given some relaxation then students took levity and stopped taking classes by giving false excuses. Many colleges around the nation have now fully digitalised their classrooms, understanding the needs of the hour. In the midst of this ambiguity and chaos, alternative modes are introduced to comfort the teachers and learners. Teaching contents and hours are made easier. For the seamless working of the education institutes, curriculum, schedules, number of tests, and assignments are changed and given weightage over traditional exams. Face-to-face feedback forms are modified and improved to gather the information for recommendations. Many students, in the beginning, complained about the slow internet facility at home, a few not having proper gadgets, many facing disturbances in the domestic environment, and even a few students find it really frustrating and challenging to focus. For many years, they are accustomed to the traditional mode of teaching. They prefer the natural environment of the classroom amongst their friends to study and chill out but this crisis has no doubt taken that freedom from them. Although, more emphasis is given to students' requirements, and lessons are tailored for easier online delivery. Therefore, it is recommended to have further studies and research in this area to provide the data and findings on effectiveness of online teaching-learning.

References

All India Survey on Higher Education (AISHE). (2019). Ministry of Human Resource Development, New Delhi: Government of India.

Barboni, L. (2019). From shifting earth to shifting paradigms: How webex helped our university overcome an earthquake. *CISCO, Upshot By Influitive*..

Dhawan, S. (2020). Online learning: A panacea in the time of Covid-19 crisis. *Journal of educational technology systems*, *49*(1), 5-22.

Martin, A. (2020). How to optimise online learning in the age of coronavirus (Covid-19): A 5-point guide for educators.

Schneider, S. L., & Council, M. L. (2021). Distance learning in the era of Covid-19. *Archives of dermatological research*, *313*(5), 389-390.

Singh, V., & Thurman, A. (2019). How many ways can we define online learning? A systematic literature review of definitions of online learning (1988-2018). *American Journal of Distance Education*, *33*(4), 289-306.

Tsitsia, B. Y. (2020). Assessing Teacher-Trainees' Perceptions Regarding the Online teaching-learning mode of the Agricultural Science Course. *Int. J. Educ. Res*, *8*, 111-124.

UGC (2020),. April). Report of the UGC committee on examinations and academic calendar for the universities in view of Covid-19 pandemic and its subsequent lockdown. Retrieved from https://www.ugc.ac.in/pdfnews/4276446_UGCGuidelines-on-Examinations-and-Academic-Calendar.pdf

WHO Coronavirus Disease (Covid-19) advice for the public https://www.who.int/emergencies/diseases/novelcoronavirus-2019/advice-for-public

Williams, K.C., Morgan, K., & Cameron, B.A. (2011). How do students define their roles and responsibilities in online learning group projects?. *Distance Education*, *32*(1), 49-62.

9

Indian Higher Education Landscape with Reference to SDG 4

R. Shital and *Ramesh Annemalla*

Introduction

Transforming our world: the 2030 Agenda for Sustainable Development was recommended by member states at the United Nations 70th session of the General Assembly in September 2015. The well-being of both people and the environment is supported by the concept of sustainable development (UNESCO, 2016a). In contrast to the Millennial Development Goals of 2000, which focused mostly on poor countries, Agenda 2030 has a global scope. The SDGs emerged from the Millennium Development Goals in the year 2016 comprises of 17 goals related to poverty, hunger, gender equity, education, and so on. Goal 4 on education focuses on ensuring that all people have access to high-quality education and compelling nations to "encourage opportunities for lifelong learning for everyone" (United Nations, 2015). It lays out a normative and aspirational vision of a better future that depends on collaboration and interdependence. This goal outlines the targets and objectives of the global education community for 2030 from the perspective of lifelong learning. All learning activities carried out throughout life with the goal of enhancing knowledge, skills, and competences from a personal, civic, social, and employment-related viewpoint are referred to as lifelong learning (UIL, 2015).The fact that more and more people are enrolling in colleges and universities after completing secondary school is one of the strong reasons to make higher education the aim of Goal 4, especially the target 4.3 (Owens, 2017).

Education is one of the significant determinants of the human resource development which allows people to get better insights into the complex world, enabling them to harness and experience their potential capabilities (Radhakrishna, 2020). In the context of globalization and fast-changing technologies along with emerging knowledge-based economies, higher and technical/professional education has a significant role to play (Monteiro et al. 2020).The study of enrolment in higher education of various countries shows that there is a positive association between per capita GDP and the gross enrolment ratio (GER) in higher education of the countries (Agarwal, 2006).This makes sense because a critical component of higher education is the transfer of practical and employable skills. Higher education not only prepares individuals to fill a variety of economic responsibilities in the society but also promotes technical advancement, thereby contributing towards economic growth (Agarwal, 2007; Pellini & Bredenberg, 2015).

Higher education enrolment has been rising gradually across the globe. The central govt. of India has been taking several steps and measures for improving the higher education system in our country. Even globally, various advanced economies like USA, UK and China have implemented various strategies and plans for the growth of higher education as they believed that higher education is the strategic driving force for sustainable economic growth in their countries (Agarwal, 2006). India stands third in terms of student population in higher education in the world next to the United States and China (Tilak, 2018; Singh, 2011) in terms of enrolment and the no. of institutions (Mittal et al. 2020). This paper aims to briefly examine the higher education in India – its evolution and growth for the period 2010-11 to 2019-20 and also to examine the role of higher education towards attaining the SDG-4 target 4.3.

Glance on SDG-4

One of the first attempts to conceptualise sustainable development was the United Nations Conference on the Human Environment in Stockholm in 1972, which stressed that communities must strike a balance between the need to improve one's quality of life and the need to continue the sustainable exploitation of natural resources (United Nations, 1972). Sustainable Development indicate a different phase of globalization in achieving this action plan for people, the environment, and prosperity (Eurostat, 2017). For the first time in human history, all nations have concurred on a set of time-bound goals for sustainable development (Sachs et al. 2019). The agenda 2030 of SDGs facilitate nations to integrate social, economic and their ecological dimensions. The goal 4 strongly emphasises on inclusive, equitable and quality education and chances for lifelong learning for all through the accomplishment of the 10 targets.

SDG 4 is one of the most significant goals out of the 17 goals as it serves as the foundation for all the other goals to be accomplished. Ten targets in SDG 4 are monitored by 11 indicators. The seven "outcome-based targets" are: universal literacy and numeracy; free primary and secondary education; universal access to high-quality early childhood education; cost effective technical, vocational, and higher education; large availability of people with skills necessary for financial security; elimination of all forms of discrimination in education; and education for sustainable development and global citizenship. The three "ways of accomplishing targets" are to construct and improve inclusive and secure schools, raise the number of higher education scholarships available to developing nations, and boost the availability of competent teachers in such nations.

> "*Target 4.3 specifically refers to equality of access amongst men and women to technical, vocational and tertiary level education, including university*" (UNESCO, 2017).

Further, some SDG 4 targets specifically call for university action, while others are crucial for learning and teaching in HEIs (Leal Filho et al. 2019a). Such a goal aims to ensure that higher education should not only be accessible to both men and women but also should be of highest quality. Lifelong learning also incorporates various, flexible learning routes, entry and re-entry points for all ages, and increased linkages between formal and non-formal structures, including formal validation of knowledge, skills, and competences earned through technical and vocational education (Owens,). As a result, including these dimensions in attaining these goals will produce effective results.

Growth of higher education in India

The British government laid the groundwork for India's contemporary higher education system in the middle of the nineteenth century. The primary institutions of Indian higher education in India comprise of universities and its constituent colleges. India's first three universities were established in the then Bombay, Calcutta, and Madras in the year 1857 (Jayaram, 2006). These universities had 27 colleges affiliated to them. The number of universities rose to 19 and several colleges were established by the time India attained independence (CABE, 2005a). In an effort to be viewed as a welfare state during the 1960s and 1970s, the government not only financed the higher education by establishing colleges and universities but also assumed financial control of the private colleges (Agarwal, 2007). Such institutions were treated as private aided institutions and/or grant-in-aid institutions. However, public funding included lot of regulations, thereby reducing the autonomy of these institutions. According to Franco et al. (2018), institutional role is "absolutely important," and they also have a "social obligation" to promote sustainable development skills among their student bodies and communities. Until 1980s, the expansion of higher education was mostly restricted to courses in humanities and languages, except a few institutes offering professional/technical education such as the Indian Institutes of Technology (IITs) and National Institutes of Technology for engineering and the Indian

Institutes of Management (IIMs) for management education. The setting up of IITs, NITs and IIMs has boosted the higher education in India. The first IIT was established in Kharagpur, West Bengal, in the year 1951 and there are total of 23 IITs set up in the period of 1951-2016 in different parts of the nation. Two IIMs were established in the year 1961 at Calcutta and Ahmedabad and thereafter 20 IIMs came into existence in the period of 1961-2016. These institutions played a significant role in shaping the higher education system of the country to attain sustainable growth. These institutions help in incorporating sustainability into their curricula and also community outreach initiatives for years (Leal Filho et al. 2019b).The main objective of any educational institution should be to contribute towards the progress of the country by creating a strong knowledge pool and offering employment opportunities for the youth (Dhingra & Kundu, 2021).

The following table gives a glance of the growth of the number of universities and institutions for a decade ranging from 2010-2011 to 2019-2020 depicting the growth and expansion of higher education in India:

Table 1: Progress of Higher Education in India

Year	Universities	Colleges
2010-11	621	32974
2011-12	642	34852
2012-13	667	35525
2013-14	723	36634
2014-15	760	38498
2015-16	799	39071
2016-17	864	40026
2017-18	903	39050
2018-19	993	39931
2019-20	1043	42343

Source: AISHE Reports 2016-17 & 2019-20.

India's higher education system, at the time of its independence was limited only to the rich and elite. But due to the expansion of the higher education system in our country, it considerably reached a broader population. As of now, there are 37.4 million students joined in higher education at 1043 universities and 42,343 colleges, increasing from 0.2 million students and as many as 20 universities in the year 1950-1951 (AISHE 2019). The table above depict that with the emergence of SDG-4 has paved the way to the rise in the no. of higher education institutions in terms of new universities and colleges all over the country. This has widened the opportunities for all the people completing their secondary level of education to opt for higher education for better employment and career prospects.

Way towards achieving SDG 4 – target 4.3

Globally, HEIs are taking the lead in creating and putting into action sustainable practises and projects (Berchin et al. 2021). The expansion in terms of no. of institutions alone is not sufficient for achieving the goals, but accessibility of these institutions in terms of enrolment is a matter of concern.This accessibility should provide equal opportunities to all the socio-economic groups irrespective of gender. The GER in higher education of a country is influenced by a number of factors, including secondary school enrolment, retention, and completion (Mittal et al. 2020). As more students finish their secondary schooling, they turn to higher education to assist them get ready for new professional paths (Owens, 2017). According to a UNESCO estimate from 2017, "between 2000 and 2014, the number of students in higher education institutions more than doubled, going from 100 million to 207 million". This indicates that the no. of people enrolling in higher education globally is increasing exponentially (p. 1). The following graph depicts the GER in higher education in India for the period 2010-11 to 2019-20:

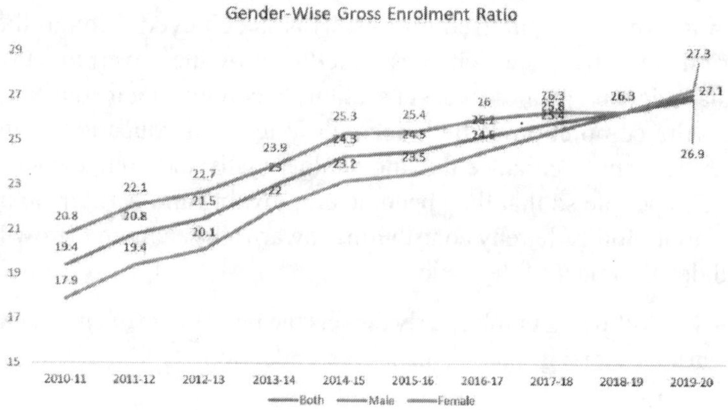

Fig. 1: Gender-wise Gross Enrolment Ratio.

Source: AISHE Reports 2016-17 & 2019-20.

Gross Enrolment Ratio (GER) in Higher education calculated for 18-23 years of age was 19.4 percent in 2010-11. In the year 2019-20, it increased to 27.1 percent. Also because of the gender discrimination, highly prevalent in India, males are given more preference for education than females, resulting in a weak female students' enrolment in higher education (Dominic & Fulgence, 2019). However, the government's attention on girls' education and socially backward class empowerment has improved the enrolment and participation of women, SCs, and STs in higher education, female GER in higher education is improving (AISHE, 2019). The male GER is 20.8 percent and that of females is 17.9 percent in the year 2010-11 and 26.9 percent and 27.3 percent respectively in the year 2019-20. These figures clearly depicts that the Indian higher education system is moving forward towards the attainment of SDG-4.

Even the enrolment in higher education of various social groups is another important determinant that depicts the achievement of the target 4.3 of the SDG. Over the decade period, it is observed that the enrolment of the weaker sections of the society has been improving. This clearly indicates that the expansion of higher education in our country is not just limited to the rich and elite but also to other marginalised groups. This is achieved through the various initiatives and schemes undertaken by the government of India to encourage these weaker sections to continue their education after the completion of their secondary level of education. This enables them to enhance the knowledge, skills and competencies of such people so that they become employable and participate in the labour force, thereby contributing towards the economic growth and development of the nation.

The following graph clearly depicts the percentage of enrolment of various social groups in the higher education:

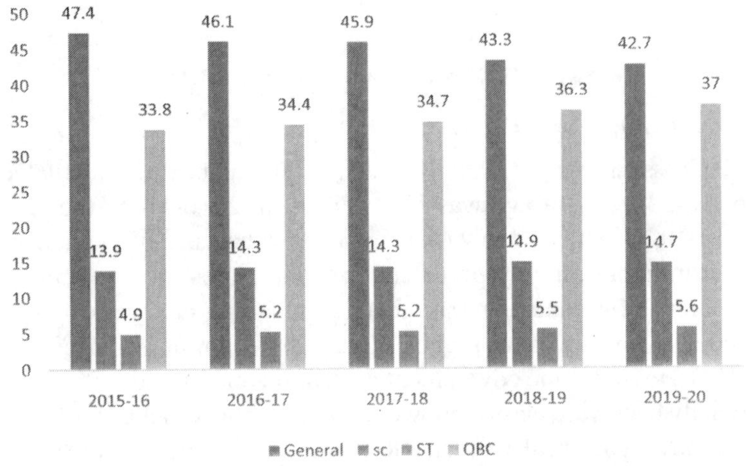

Fig. 2: Enrolment among various Social Groups.

Source: AISHE Report 2019-20.

In the year 2015-16, the enrolment of SC category of students in higher education is 13.9 percent but by the end of 2019-20, it is 14.7 percent. While the STs rose to 5.6 percent in 2019-20 from 4.9 percent in the year 2015-16. Even the other backward classes enrolment has improved from 33.8 percent in 2015-16 to 37 percent in 2019-20.

This is also evident in some of the major states of India such as Tamil Nadu, Kerala, Andhra Pradesh, Telangana, Karnataka etc. The GER of these states is in increasing trend if we consider the data pertaining to five years ranging from 2015-16 to 2019-20. This improvement can be rightly pointed towards the initiatives and measures taken up by both the central and their respective state governments. National scholarships and fellowships for undertaking research and innovation, special scholarships for socially backward and persons with disabilities are some of the major initiatives taken up by the central government towards the improvement of higher education in the country and thereby attaining the target 4.3 of SDG 4.

The table below shows the GER of some of the major states of India: -

Table 2: GER of major states of India during the last five years

States	2015-16	2016-17	2017-18	2018-19	2019-20
Tamil Nadu	44.3	46.9	48.6	49	51.4
Telangana	36.3	35.8	35.7	36.2	35.6
Uttara Khand	33.3	33.4	36.3	39.1	41.5
Himachal Pradesh	32.5	36.7	37.9	39.6	40.8
Andhra Pradesh	30.8	32.4	30.9	32.4	35.2
Kerala	30.8	34.2	36.2	37	38.8
Maharashtra	29.9	30.2	31.1	32	32.3
Arunachal Pradesh	28.7	28.9	29.7	29.7	35.4
Karnataka	26.1	26.5	27.8	28.8	32
Haryana	26.1	29	28.7	29.2	29.3
India	24.5	25.2	25.8	26.3	27.1
Gujrat	20.7	20.2	20.1	20.4	21.3
Rajasthan	20.2	20.5	21.7	23	24.1
Madhya Pradesh	19.6	20	21.2	21.5	24.2
West Bengal	17.7	18.5	18.7	19.3	19.9
Bihar	14.3	14.4	13	13.6	14.5

Source: AISHE Reports 2016-17 & 2019-20.

The graph below shows the top 5 and bottom 5 Indian states with respect to GER for the period 2015-16 to 2019-20: -

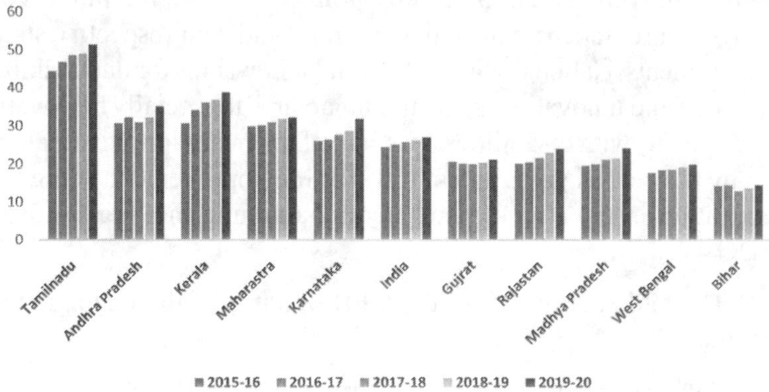

Fig. 3: GER of India and its major states.

Source: AISHE Reports 2019-20.

It is clear from the above graph that Tamil Nadu, Andhra Pradesh, Kerala, Maharashtra and Karnataka are better performing states in terms of GER. Their enrolment in higher education is more than that of the national average which is 25.78 considering the last five years. The factors determining the better performance of these top performing states are better infrastructure, affordable higher education, quality of teachers and better coordination and determination of the state and the central governments. While Gujarat, Rajasthan, Madhya Pradesh, West Bengal and Bihar are the poor performing states with respect to higher education in terms of GER. The reasons for poor performance of these states are lack of resources, poor quality of infrastructure and teachers, lack of research and innovation etc. However, these states can improve their performance in higher education by taking steps for providing better quality education through better resources and infrastructure.

Conclusion

The overall conclusion of the study is that, in order to attainable growth in any economy higher education place an important role. The SDG-4 aiming for inclusive and equal access to higher education is successful in encouraging a greater number of people, irrespective of gender and socio-economic status, to continue their higher education after the completion of secondary level of education. This enables them to not only improve their standard of living but also contributing towards the economic prosperity. Higher education institutions (HEIs) have become change agents as a result of the growing society need for more sustainable practices. The government of India taking several initiatives and strategies to increase the GER in higher education. The study therefore concludes that the exponential growth of higher education in India, increasing overall GER and also gender-wise enrolment implies that India is moving towards the achievement of SDG-4 especially target 4.3. The GER indicates that the enrolment women in higher education in India has surpassed that of men. We can say that India has achieved the objective of gender equity. Additionally, social and economic justice has improved and more no. of students from underprivileged sectors are enrolled in higher education. While some efforts are being made to ensure quality parameters through various skill development programmes, much more has to be done. Multidisciplinary approaches should be used in the HEIs to develop the learning environment and curricula, ensuring that sustainability studies are integrated into all academic fields and courses.

Further scope

The study is confined to target 4.3 of the SDG-4. However further study can be undertaken on other targets of the SDG-4.

References

Agarwal, P. (2007). Higher education in India: Growth, concerns and change agenda. *Higher Education Quarterly, 61*(2), 197–207.

Agarwal, Pawan (2006) : Higher Education in India: The Need for Change, Working Paper, no. 180, *Indian Council for Research on International Economic Relations (ICRIER)*, New Delhi.

AISHE Reports, various years, Department of Higher Education, MHRD, New Delhi.

Berchin, I. I., de Aguiar Dutra, A. R., & Guerra, J. B. S. O. D. A. (2021). How do higher education institutions promote sustainable development? A literature review. *Sustainable Development*, 29(6), 1204-1222.

CABE Committee. (2005a). *Report of the Central Advisory Board of Education (CABE)* Committee on Autonomy of Higher Education Institutions. Government of India. June, 2005.

Dhingra, M., & Kundu, S. C. (2021). Factors affecting placement and hiring decisions: A study of students' perceptions. *Industry and Higher Education, 35*(3), 223–232.

Eurostat. *Sustainable Development in the European Union—Monitoring Report on Progress towards the SDGs in an EU Context*; Publications Office of the European Union: Luxembourg, 2017.

Jayaram, N. (2006). National perspectives: India. In: Forest, James J. F. and Altbach, Philip G. (eds.). *International Handbook of Higher Education*. Springer. Dodrecht, The Netherlands.

Leal Filho, W., Shiel, C., Paço, A., Mifsud, M., Ávila, L. V., Brandli, L. L., ... & Caeiro, S. (2019). Sustainable Development Goals and sustainability teaching at universities: Falling behind or getting ahead of the pack?. *Journal of Cleaner Production, 232*, 285-294.

Leal Filho, W., Vargas, V. R., Salvia, A. L., Brandli, L. L., Pallant, E., Klavins, M., ... & Vaccari, M. (2019). The role of higher education institutions in sustainability initiatives at the local level. *Journal of Cleaner Production, 233*, 1004-1015.

Mittal, P., Radkar, A., Kurup, A., Kharola, A., & Patwardhan, B. (2020). Measuring access, quality and relevance in higher education. *Economic and Political Weekly, 55*(24), 34–38.

Monteiro, S., Almeida, L., Gomes, C., & Sinval, J. (2020). Employability profiles of higher education graduates: a person-oriented approach. *Studies in Higher Education, 0*(0), 1–14.

Owens, T. L. (2017). Higher education in the sustainable development goals framework. *European Journal of Education, 52*(4), 414–420.

Pellini, A., and Bredenberg, K. (2015). Basic education clusters in Cambodia: looking at the future while learning from the past. *Development in Practice, 25*(3), 419–432.

Radhakrishna, R. (2020). Recent papers of interest. Towards reforming the education system in India. *Indian Public Policy Review.* Vol. 1(2).

Sachs, J., Schmidt-Traub, G., Kroll, C., Lafortune, G. and Fuller, G. (2019), "Sustainable Development Report 2019", *Bertelsmann Stiftung and Sustainable Development Solutions Network (SDSN)*, New York, NY.

Singh, J. (2011). Higher Education in India–Issues, Challenges and Suggestions. *Higher Education*, September, 93-103.

Tilak, J.B.G (2003). Vocational Education and Training in Asia. *International Handbook of Educational Research in the Asia-Pacific Region*, Vol.11, pp. 673-686.

UIL. (2015). Conceptions and realities of lifelong learning (Background paper for Global Education Monitoring Report 2016). Paris, France: UNESCO.

UNESCO Institute for Statistics (2019), "SDG 4 data digest: how to produce and use the global and thematic education indicators".

United Nations. (1972). Report of the United Nations conference on the human environment.

10

Achieving Sustainable Sanitation Coverage in Urban Slum Settings

J.V. Arun and *A. Premkumar*

Introduction

Among the world nations, India is known for its vivid culture portrayed with flamboyance, knowledge and ethical values. It is a nation with citizens who value dignity more than their life. However, significant gap also exists between knowledge and practice which is evident from unhealthy practice of defecating in the public. Even in the advanced era where people are busy propagating importance on human well-being, open defecation has remained as an indelible blot in the Indian society. It is not just well-established traditional practice that can be seen as a problem but also due to the existence of poverty that forces people not to prioritise toilets. India's massive open defecation problem is holding back the progress on sanitation. Access to basic sanitation services is one of the priority areas of action of the Indian government. But their slow progress on sanitation is a major concern as it affects health and productivity (Cullet et al. 2019; Vyas et al. 2019 and Winter et al. 2019). The role of sanitation programmes is crucial for addressing these challenges and achieving clean India. Since the inception of National Water Supply and Sanitation Programme in the year 1954, there are significant changes in the formulation and implementation of the sanitation programmes. Many new programmes were initiated and defunct programmes were revised with new goals and targets. However, previous programmes like Central Rural Sanitation Programme (CRSP), Total Sanitation Campaign (TSC), and Nirmal Bharat Abhiyan (NBA) fell short of their target of achieving a clean India. Consequently, the Government of India (GoI) launched Swachh Bharat Mission (SBM) on 2nd October 2014; with two sub–missions, the SBM–Rural and the SBM–Urban. It plans to make the nation Open Defecation Free (ODF) by means of constructing individual household, cluster and community toilets. The flagship programme is being put into operation by the Ministry of Urban Development and by the Ministry of Drinking Water and Sanitation. As envisaged, India has been affirmed as Open Defecation – Free by the Government on 2nd October 2019.

All stakeholders should recognise that there is a need to ensure sustainability of the ODF status and for which achieving ODF + and ODF ++ is a necessity as approved by the Ministry of Housing and Urban Affairs. These two are nothing but up-gradation status of ODF with additional requirements. Ensuring proper maintenance of toilet facilities is referred as SBM ODF+, and safe collection, treatment and disposal of faecal sludge and sewage are referred as SBM ODF++. Safe sustainable sanitation for all is achievable only through longer-term process of behavioural change. Encouraging all stakeholders to be an inherent part of the post-ODF strategies and creating behavioural change awareness would help the country to achieve sustainable sanitation. In this setting, the paper makes an effort to review obstacles that threaten our ability to expand sanitation coverage to meet growing needs in urban areas. It is necessary to study the presence of sustainable sanitation which is understood as solutions that, apart from protecting and promoting human health that are "economically viable, socially acceptable, and technically and institutionally appropriate and those that protect the environment and natural resources".

Sanitation target of the Sustainable Development Goals

Sanitation is fundamental to healthy and productive urban life, and the provision of sanitation services for fast-growing urban population happens to be one of the world's most urgent challenges. Meeting sanitation needs means developing systems that are safe, affordable and functional, as well as sustainable over the long term. This requires a better understanding of the dysfunctions, discontinuities and inequalities in the current models of sanitation provision in the urban areas. Clean water and sanitation are one of the 17 Sustainable Development Goals (SDGs) adopted by the UN Member States at the Sustainable Development Summit, 2015. Sanitation is a major component of SDGs, of which Target #2 of Goal #6 is to achieve adequate sanitation for all (Figure-1). But, poor sanitation behaviour/attitude exhibited in India are going to be an obstacle to achieve that target by 2030. India, as a signatory to the SDGs needs to work out its priorities to achieve those milestones. UNICEF is working with the Indian government to perk up the efficiency of the government's sanitation programme, SBM.

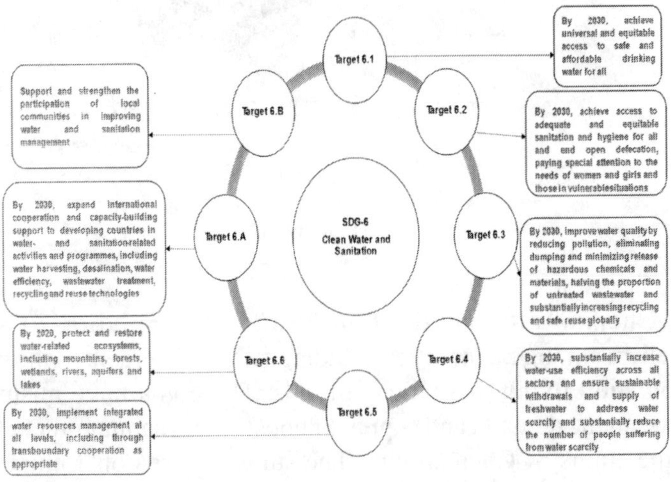

Fig. 1: SDG-6 Clean Water and Sanitation

Source: Developed by authors.

UNICEF also provides technical support to state governments to improve the situation of WASH (water, sanitation and hygiene) in schools and in health centres.

3. Methodology

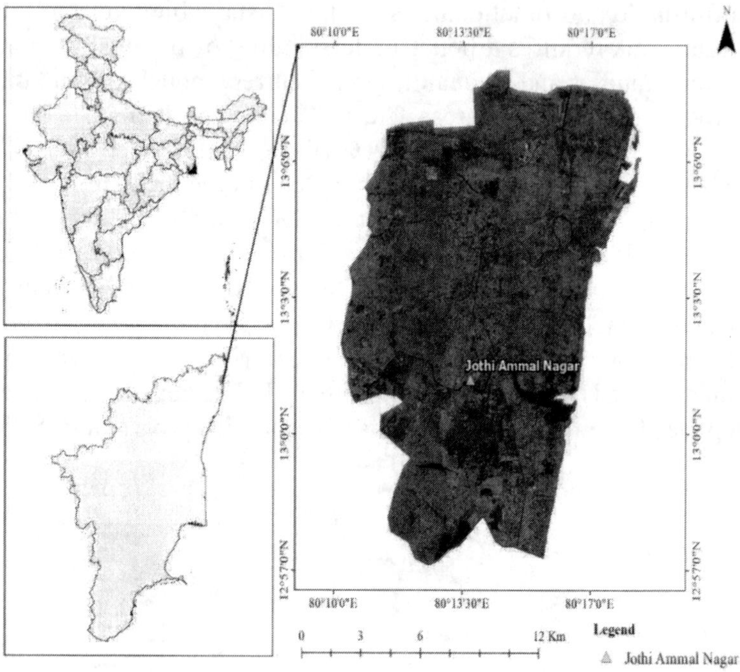

Fig. 2: Study Area

Source: Developed by authors.

Chennai city has the fourth highest population of slum dwellers among major cities in India, with 18.6 percent of the city's population living in slum conditions. There is a huge mismatch between the demand and supply of housing and sanitation facilities in the slums of Chennai city. The study focuses on Jothiammal Nagar (ward no. 171), which is located in the Saidapet area of Chennai Corporation. Figure 2 gives the location of Jothiammal Nagar which lies in the Adyar zone. There are approximately 600 households with a population of 2520 (District census handbook, 2011). The area is affected by garbage and sewage problem. Further, people do not have any proper sanitation facilities. This formed the motivation for choosing this as study area. A total of 100 households were surveyed from the study area and it was carried out in May – June, 2021. Table – 1 shows the characteristics of the sample population and the households include all categories of urban population.

Table 1: Characteristics of the sample population

Total Households	**N=100**	
Population	Total	393
	Male	189
	Female	204
Religion	Hindu	94
	Christian	4
	Muslim	2
Community	BC	19
	MBC	30
	SC	51
	Total	100
Total Literates	Male	146
	Female	160
Dependent Members in Households	Male	75
	Female	167
Total Income Earners		151
Households with Latrine Facilities		88
Households without Latrine Facilities		12

Source: Primary Survey, 2021.

4. Factor Analysis

As the survey covers large set of measured variables, factor analysis was used to reduce the dimension of the original variables by constructing relatively few categories. This technique of grouping variables into factors is helpful to determine the patterns of relationship among the observed data. SPSS was used to perform factor analysis of the data collected from the survey forms. To examine the appropriateness of factor analysis, Kaiser-Meyer-Olkin (KMO) and Bartlett's Test measure of sampling adequacy was used. Then Communality, Eigen values and Sum of squared loadings were obtained. For results, rotation technique was adopted which reveal different structures of the data.

Table 2: Descriptive Statistics and Communalities

Sl. No	Variables	Mean	Std. Deviation	Analysis N	Extraction
1	Age of the household head	44.2600	9.54164	100	.889
2	Sex of the household head	.9000	.30151	100	.586
3	Educational Level of the household head	2.2100	1.17461	100	.628
4	Occupation of the head	3.6800	1.49666	100	.566
5	Family Size in Numbers	3.9500	1.03840	100	.566
6	Monthly income of the family	10395.0000	3454.82754	100	.737
7	Source of water used for non-drinking purpose	3.40	.964	100	.795
8	Satisfied with the water supply	1.7700	.42295	100	.737
9	Having Toilet	1.1200	.32660	100	.956
10	Functional status of toilet	1.0100	.50242	100	.899
11	Reason for not functioning of toilet –Dirty	.2400	.45216	100	.821

Sl. No	Variables	Mean	Std. Deviation	Analysis N	Extraction
12	Functionality of public toilets	1.9200	.27266	100	.468
13	Where does excreta discharge from the toilet into/ to?	1.9400	1.70454	100	.914
14	How many times the toilets are cleaned in a week?	2.1600	.52647	100	.678

Source: Computed Data.

Table 2 reveals the descriptive statistics and the communalities of the variables. In particular, the mean, standard deviation and the extracted value of the respondents' response were taken into account. The outcome of the mean value depicts, 'Monthly income of the family' and 'Age of the household head' influences all the other variables. This is due to the fact that those two variables have the highest mean value, i.e., 10395.0 and 44.26. And, the extraction column of the communalities provides information about the degree of variance in each variable which can be explained by the retained factors.

The KMO test is used to measure the sampling adequacy of the collected data. The thumb rule of this test is that the value closer to 0.5 is acceptable. Moreover, the outcome of KMO measure (KMO = 0.547; p-value < 0.00) shows that the data is adequate to run the dimension reduction analysis. Further, Bartlett's test is used to measure the strength of the association among the variables. The value of Bartlett's test suggests that, it is statistically significant at one percent level of significance ($BT_{(105)}$= 855.109; p-value < 0.00). This significant value specifies that relationship matrix isn't an identity matrix (Table 3). Hence, factor analysis is considered suitable for further analysis of the dataset.

Table 3: KMO and Bartlett's Test for adequacy of data

S. No	KMO and Bartlett's Test		P-Value
1	Kaiser-Meyer-Olkin Measure of Sampling Adequacy.		.547
2	Bartlett's Test of Sphericity	Approx. Chi-Square	855.109
		Df	105
		Sig.	.000

Source: Computed Data.

Eigen value shows the number of extracted factors which are exposed to the factor analysis. In Table 4 extracted sums of squared loadings results are used to explain the significant and insignificant factors. Factors having eigen values greater than one was considered significant and all others were considered insignificant and disregarded. There are six factors each having eigen value exceeding one. Hence, six factors were extracted for analytical purpose and the percentages of variance explained by factors one to six are 24.876, 14.785, 11.777, 8.100, 7.526 and 7.025 respectively. The six–factor solution accounted for 74.09 percent of the total variance and all the other remaining factors were not significant.

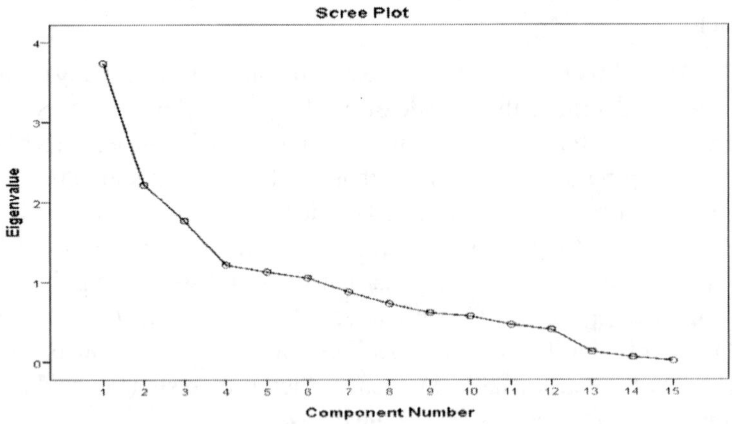

Fig. 3: Scree plot graph.

The Eigen values of all the factors are explained with a scree plot graph (Fig. 3). Under this criterion, values which are less than 1 are not considered for further analysis. This is a plot wherein the values after 6th factor were less than the Eigen value (< 1). Therefore, the curve starts to flatten between the 6th and 7th factor, implying that each successive factor accounts for smaller and smaller amounts of the total variance.

Table 4: Outcome of Eigen values

Initial Eigen values			**Extraction Sums of Squared Loadings**		
Total	**% of Variance**	**Cumulative %**	**Total**	**% of Variance**	**Cumulative %**
3.731	24.876	24.876	3.731	24.876	24.876
2.218	14.785	39.661	2.218	14.785	39.661
1.767	11.777	51.438	1.767	11.777	51.438
1.215	8.100	59.538	1.215	8.100	59.538
1.129	7.526	67.064	1.129	7.526	67.064
1.054	7.025	74.090	1.054	7.025	74.090
.877	5.845	79.935			
.733	4.886	84.821			
.618	4.117	88.938			
.577	3.848	92.786			
.471	3.140	95.926			
.409	2.727	98.653			
.125	.837	99.490			
.062	.416	99.906			
.014	.094	100.000			

Source: Computed Data.

Rotated Component Matrix is arrived at after performing factor rotation. This type of matrix represents the pattern of loadings of the original indicators on extracted factors. The dimension reduction process illustrates that, all the factors were interrelated with one another. Six new factors are successfully constructed using factor analysis and assigned as the factors affecting sanitation behaviour of the urban slum dwellers (Table 5). The six factors are labelled according to their nature as; (1) accessibility of toilets, (2) economic status, (3) condition of toilets, (4) family size, (5) availability of water and (6) users age. Three items are factored in toilet accessibility with factor loadings ranging from 0.69 to 0.93 and wherein two items are dropped because of cross and low loading. Similarly, economic status group comprised of three items and item S.No.14 is dropped due to cross loading. On the other hand, both condition of toilets and family size are factored with two items. Finally, availability of water and users age comprised of one item, each. The factor loadings are 0.66 and 0. 91 respectively. Among these core factors, items factored in accessibility of toilets shows the highest percentage of variance.

Table 5: Rotated Component Matrix

Sl. No	Variables	1	2	3	4	5	6
1	Age of the household head						.910
2	Sex of the household head				-.564		
3	Educational Level of the household head		.707				
4	Occupation of the head		.619				
5	Family Size in Numbers				.600		
6	Monthly income of the family		.657				
7	Source of water used for non-drinking purpose				.578	-.648	
8	Satisfied with the water supply					.662	
9	Having Toilet	.931					
10	Functional status of toilet	-.546		.648			
11	Reason for not functioning of toilet – Dirty	.691					

Sl. No	Variables	1	2	3	4	5	6
12	Functionality of public toilets	-.642					
13	Where do excreta discharge from the toilet into/to?	.865					
14	How many times the toilets are cleaned in a week?		.561	.504			

Source: Computed Data.

Discussion of results

Accessibility of toilets

Water and Sanitation Programme's (2011) report pointed out that inadequate sanitation causes the poor in India to endure considerable adverse bearings on their lives, health, and limited financial resources. A study in Chennai found that women without access to toilets restricted their urination, whereas women with access to toilets urinated as frequently as men (Diverde, 2013). It has been sufficiently established that one of the toughest daily experiences for adolescent girls in India is the experience of going out to relieve themselves, be it to community toilets or to open defecation spots. Many of the women who were restricting urination because of lack of access described symptoms associated with urinary tract infections, for which many of them frequently had to take medications. Improving access to the toilets is one of the keys to achieve SDG, as one of the targets of Goal 6 of SDGs is to "achieve access to adequate and equitable sanitation and hygiene for all and end open defecation, paying special attention to the needs of women and girls and those in vulnerable situations by 2030".

Economic status

A look at the economic status is likely to reveal preference for using toilets instead of open defecation, there is a positive association between economic well-being and quality of toilet facilities. Brindha Vijayalatha (2013) made an assessment on the impact of poverty on Hygiene and Sanitary issues among women in Tamil Nadu with reference to most backward, SC/ST communities. The study found out marginalised women suffer the most from various health issues due to poor sanitation, 75 percent of girls experienced some kind of problem associated with menstruation. Sandeep and Pradeep (2019) suggested that government should give more subsides to construct toilets in every household and also school children–led total sanitation (SLTS) model of Nepal should be adopted which may be helpful to inculcate healthy sanitation practices.

Condition of toilets

Poor sanitary conditions and inadequate waste disposal systems in crowded urban and peri-urban areas is the important reason for the spread of infectious diseases like tuberculosis, pneumonia, and diarrhea. This had contributed to the high mortality and morbidity among children in poor households in Asia, Africa, and Latin America (Saikia, 2014). The cause of many of our diseases is the condition of our lavatories and therefore, it is an absolute necessity to have of a clean place for answering the call of nature. A study showed that efficient sewerage systems typically reduce diarrhea incidence by about 30 percent or perhaps as much as 60 percent (Norman et al. 2010). The presence of efficient drainage and sewerage system is a major factor in the prevention of spread of waterborne diseases in major cities. Moreover, awareness regarding cleanliness of the toilets to be created among the urban poor in order to gain good health (Hazarika, 2015).

Family size

There is evidence from research on sanitation that links lack of adequate sanitation in urban slums to household size. Issaka et al. (2019) made a mixed–method study to determine the factors that influence open defecation. The study has identified that education, household size, occupation, income, traditional norms, beliefs and ownership are the six important factors that affect an individual's sanitation practices. It was also noted in Panchang (2019) that household size, household water sources, and avoidance of open defecation are significantly associated with interest in usage of toilet.

Availability of water

The urban poor lack adequate access to services like water supply, sewerage, low-cost housing, education and public health. The population living in the slum settlements in particular have been affected most adversely by uncertain access to water (Steinmann et al. 2015; Manasi et al. 2017); mainly women felt unsafe to access toilet facilities due to inadequate water supply (Madhusudan and Parag, 2018). Evidences show that dependence on traditional water sources and temporary sources that usually dry up in summer, impacts the ability to use toilets. The seasonal variations in toilet use however could be checked if water supply is ensured for each and every household. This will also have an impact on the sustainability of toilet use. Appropriate measures should be taken by government to increase the awareness among public about water management to make positive impact on toilet usage (Yogananth and Tarun, 2018).

Users age

Different sections of population have special needs for access to sanitation, for example the elderly, adolescent girls, marriage-age boys, children, and even infants below 2 years. The varied nature of their needs and how they can be met has to be integrated for sustainability to be achievable. Utmost care has to be taken to ensure that the specific needs of this population are addressed. Specific interventions for children below one year in the form of Potties would be helpful in habit formation of children and they became change agents for the entire family and community.

Conclusion

A key element of improving sanitation facility is eliminating open defecation. In practice, it is those who have access to a latrine who have the capacity to choose between latrine use and open defecation. Of course, these decisions are influenced by a person's behaviour. Study recommends that action-oriented programme should be followed for overall behavioural change to promote better latrine usage. Consequently, the community needs to be sensitised for awareness generation and catalysing behavioral change on individual hygiene practices, apt storage and treatment of drinking water, and safe disposal of waste. To meet up with sanitation requirements means developing systems those are secure, reasonably priced and serviceable, as well as sustainable in the long term. Achieving them needs a better perceptive of the dysfunctions, discontinuities and disparities in the existing models of sanitation proviso in a country like India.

References

Diverde, H. (2013). The sanitary situation and its health effects on women exposed to occupational heat in Chennai, India. Umea, Sweden: Umea University.

Hazarika Minakshi, P. (2015). Sanitation and Its Impact on Health: A Study in Jorhat, Assam. *International Journal of Scientific and Research Publication*, 5(10), 1-11.

Norman, G., Pedley, S., Takkouche, B. (2010). Effects of sewerage on diarrhoea and enteric infections: A systematic review and meta-analysis. *Lancet Infect*, 10(8), 536-544.

Osumanu Issaka Kanton, Kosoe Enoch Akwasi, Ategeeng Frank (2019). Determinants of Open Defecation in the Wa Municipality of Ghana: Empirical Findings Highlighting Socio-cultural and Economic Dynamics among Households. *Journal of Environmental and Public Health*, 1-10.

Panchang, S.V. (2018). Demand for improved sanitation in an urban informal settlement in India: role of the local built environment. *International Journal of Environmental Health Research*, 29(2), 194-208.

Pandya Madhusudan N. and Shukla Parag Sunil (2018). Role of Women Led Sanitation in Community Development. *Journal of Content, Community & Communication*, 7(4), 71-77.

Philipe Cullet, Sujith Koonan and Lovleen Bullar (2019). *The Right to Sanitation in India,* Oxford University Press, 75-111.

Saikia, B. (2014). Health Challenges of Cities in Developing Countries: A Study of Guwahati City in Northeast India. *International Conference on Trends in Economics, Humanities and Management*, 1-6.

Sehsaiah, Manasi, Nagesh, L, Ramesh, H. (2017). Sanitation challenges of the poor in urban and rural settings: Case studies of Bengaluru City and rural North Karnataka. Maastricht University, UN. Working Paper Series: 026, 1-39.

Sinha Sandeep Kumar and Chaudhry, Pradeep. (2019). Tackling Open Defecation and Improved Sanitation in Developing Countries: A Toilet Talk from the State of Bihar, India.*Current Research Journal of Social Sciences*, 2(1), 47-57.

Steinmann, P., Juvekar, S., Hirve, S. and Weiss, M.G. (2015). Coping Strategies to Deal with Inadequate WASH Facilities and Related Health Risks. Research Briefing Note, Research Consortium and Water Supply and Sanitation Collaborative Council (WSSCC).

Vyas, S, Srivastav, N, Mary, D, Goel, N., Srinivasan, S., Tannirkulam, A, Ban, R, Spears, D., Coffey, D. (2019). Measuring open defecation in India using survey questions: evidence from a randomised survey experiment. *BMJ Open*, 1-9.

Winter, SC, Dreibelbis, R, Dzombo, MN, Barchi, F. (2019). A mixed-methods study of women's sanitation utilization in informal settlements in Kenya. *PLOS ONE*, 14(3), 1-20.

Yoganathan, Nallathambi and Bhatnagar, Tarun (2018). Prevalence of open defecation among households with toilets and associated factors in rural south India: an analytical cross-sectional study. *Trans R Soc Trop Med Hyg*, 1-12.

11

Socio-economic Analysis on the Impact of Covid-19 on Indian Economy: An ARDL Approach

B. Keerthika and *R. Jayapriya*

Introduction

In 2019, India experienced both health and economic shocks due to the pandemic outbreak of Covid-19. India, a densely inhabited country with limited medical resources, was obliged to enact a lockdown policy. India's economy is in decline, as is the world economy. The present pandemic is spreading throughout a fully globalised world with integrated manufacturing and financial networks. Covid-19 will have a massive influence on the Indian economy, both because of the country's lockdown, which was necessary to contain the spread of the coronavirus, and because of India's integration with the rest of the globe. No country has escaped its wrath, from developed to developing. The International Monetary Fund's chief indicated on April 9 that the year 2020 may see the worst global economic impact since the Great Depression of the 1930s, with over 170 nations expected to face negative per capita GDP growth as a result of the rampant coronavirus pandemic. According to official figures issued by the ministry of statistics and programme implementation, the Indian economy contracted by 7.3 percent in the April-June quarter of this fiscal year. This is the largest downturn since the government began recording quarterly GDP data in 1996. Following the enactment of the lockdown, it is projected that 10 million migrant workers will return to their native countries by 2020. Surprisingly, neither the state nor the federal governments had any information on the migrant workers who had lost their jobs and lives as a result of the lockdown. The epidemic has a tremendous impact on the Indian economy across sectors, in diverse circumstances of continuous, protracted, and partial lockdown, and at various degrees of capacity utilization. In an optimistic scenario, India's economy may just manage 0.5 percent positive growth in 2020, but it also confronts the risk of a 3-7 percent negative growth in worst-case scenarios (Sahoo, 2020). The shock is playing out really similar in various countries around the world in terms of demand and supply interruptions, as well as the consequent economic slowdown. However, because of the status of the economy prior to the Covid-19 period, the crisis in India may be more severe and endure longer. Following years of poor performance, India's economy had worsened dramatically by the time the first Covid-19 case was announced. As a result, Covid-19 is expected to have an economic impact on all aspects of life, including growth, global trade, financial markets, unemployment, incomes, poverty, and a number of other issues. A huge loss is expected as a result of the virus's impact on global trade. Because the Indian growth model is based on export-led growth (Mishra, 2019), a virus-related lockdown might have a substantial impact on growth. Domestic demand may be slowed due to the effects of the pandemic coronavirus Covid-19. Job losses or wage cutbacks will reduce purchasing power, and the slowing effect of deferred demand will have a longer-term impact on many industries, particularly when demand is discretionary. India's real GDP plummeted to its lowest level in more than six years in the fourth quarter of 2019-20. In 2020-21, India's growth rate is predicted to range between 5.3 percent and 5.7 percent (Patnaik, 2020).

Literature Review

The rise in new infectious disease occurrences of global importance, such as the SARS and the 2009 influenza A/H1N1 pandemic, imposes a higher necessity for interconnection between local governments and the global society. In response to this need, the WHO's International Health Regulations (IHR 2005) were changed to update monitoring capabilities and demand reporting of disease episodes that may constitute "national health emergencies of international importance". In 2000, the Global Outbreak Alert and Response Network (GOARN) was formed as a global partnership to streamline technical assistance for outbreak surveillance and response activities (Chan, E H, Brewer, T F, Madoff, L C, Pollack ,M P, Sonricker, A L., and others, 2010).

Nearly the entire world has been adversely harmed by Covid-19's uncontrolled expansion across the fringes and geologies, which has also caused significant hitch risks to the whole global economic outlook (Fernandes, 2020).

According to Barrett, R, Brown, P J. (2008), people of certain racial backgrounds and those presumed to have come into contact with the virus have become targets of societal stigma and discriminatory acts as a result of the current Covid-19 outbreak. The closure of financial markets, corporate offices, businesses, and events—which may have a significant impact on economic growth—was highlighted by Ozili & Arun (2020) and praised as a result of virus control measures.

Singh and Neog (2020) studied about the Covid-19 influence on the Indian economy by using statistical figures and economic indicators in different sectors such as trade, stock market, human capital, transportation, and travel and tourism.

Demand side actions were a major weakness in the government's economic response up until October 2020, according to Ramakumar and Kanitkar (2021), who used the Input-Output framework in their study to draw this conclusion. India's rate of GDP growth in 2020–2021 could be as low as –4.3 percent or as high as 15 percent.

The economic effects of the Covid-19's rapid global proliferation have been highlighted in a study by Karen and Preston (2020), who also speculate that it may cause a global recession.

UN assessment estimates that the Indian economy will lose $348 million as an upshot of Covid-19. According to the report, India is one of the 15 economies that have been most negatively impacted by the coronavirus pandemic (The Hindu Business Line, 2020).

Methodology and variables

The India states were categorised as Western, Eastern, Southern, Northern and North-Eastern, as follows

Region	States
Western regions	Chhattisgarh
	Gujarat
	Madhya Pradesh
	Maharashtra
	Goa
Eastern region	Bihar
	Jharkhand
	Odisha
	West Bengal
	Sikkim
Southern region	Andhra Pradesh
	Karnataka
	Kerala
	Tamil Nadu
Northern regions	Punjab
	Haryana
	Rajasthan
	Utter Pradesh
	Uttrakhand
	Himachal Pradesh
	Jammu & Kashmir

North-Eastern	Arunachal Pradesh
	Assam
	Manipur
	Meghalaya
	Mizoram
	Nagaland
	Tripura

The region-wise analysis was attempted but due to insufficient data and the result of ANOVA there is a significance difference within the states also, hence state-wise analysis was done. Using time series data, we investigate the relationship between economic growth in India during the Covid-19 crisis using the Autoregressive Distributed Lag (ARDL) model. Here, Net Domestic State Product (constant prices) is used as a proxy for the measure of economic growth from Das and Patnaik, 2020 and it is used as the independent variable. School enrollment (per gross) was taken as the control for human development, urban and rural unemployment rate, Covid deaths and inflation rate were taken as the dependent variables.

In this article, we estimate the association between financial growth and government policies, while accounting for other control variables. Regardless of whether variables are integrated, co-integrated, or not, level VARs can be tested for Granger non-causality using a direct approach created by Toda and Yamamoto in 1995. As a result, rather than its "actual" lag orders of p, a VAR is approximated with lag order (p+d), in which d is the highest possible degree of variable analysis. Then, Granger causality is examined using VAR hypothesis testing without taking into account the further lags p+1,..., and p+d. Toda & Yamamoto demonstrated how one may use conventional asymptotic theory to assess both linear and nonlinear limits in this situation. This procedure, like the ARDL technique, has recently been employed in several causality investigations. It eliminates the relatively low root and co-integration from before. The ARDL approach to co-integration has the added benefit of generating accurate long-run model estimates even when some of the indicators are uncontrolled (Odhiambo,

2009). While most traditional co-integration methods use a series of equations, the ARDL approach also uses a single reduced-form equation (Pesaran & Shin, 1999).

We estimate the association between economic growth and Covid impact by the following causality VAR empirical model using STATA v.64.

$$\Delta EGt = \alpha_0 + \sum_{t=1}^{p} \beta_{1t} \Delta EG_{t-1} + \sum_{t=1}^{p} \beta_{2t} \Delta \mathrm{SER}_{t-1} + \sum_{t=1}^{p} \beta_{3t} \Delta \mathrm{INFR}_{t-1} + \sum_{t=1}^{p} \beta_{4t} \Delta RUE_{t-1} + \sum_{t=1}^{p} \beta_{5t} \Delta \mathrm{UUE}_{t-1} + \sum_{t=1}^{p} \beta_{6t} \Delta \mathrm{COVD}_{t-1} + \mu_{1t}$$

$$\Delta \mathrm{SER}_t = \alpha_1 + \sum_{t=1}^{p} \beta_{1t} \Delta \mathrm{SER}_{t-1} + \sum_{t=1}^{p} \beta_{2t} \Delta EG_{t-1} + \sum_{t=1}^{p} \beta_{3t} \Delta \mathrm{INFR}_{t-1} + \sum_{t=1}^{p} \beta_{4t} \Delta RUE_{t-1} + \sum_{t=1}^{p} \beta_{5t} \Delta \mathrm{UUE}_{t-1} + \sum_{t=1}^{p} \beta_{6t} \Delta \mathrm{COVD}_{t-1} + \mu_{2t}$$

$$\Delta \mathrm{INFR}_t = \alpha_2 + \sum_{t=1}^{p} \beta_{1t} \Delta \mathrm{INFR}_{t-1} + \sum_{t=1}^{p} \beta_{2t} \Delta EG_{t-1} + \sum_{t=1}^{p} \beta_{3t} \Delta \mathrm{SER}_{t-1} + \sum_{t=1}^{p} \beta_{4t} \Delta RUE_{t-1} + \sum_{t=1}^{p} \beta_{5t} \Delta \mathrm{UUE}_{t-1} + \sum_{t=1}^{p} \beta_{6t} \Delta \mathrm{COVD}_{t-1} + \mu_{3t}$$

$$\Delta RUE_t = \alpha_3 + \sum_{t=1}^{p} \beta_{1t} \Delta RUE_{t-1} + \sum_{t=1}^{p} \beta_{2t} \Delta EG_{t-1} + \sum_{t=1}^{p} \beta_{3t} \Delta \mathrm{SER}_{t-1} + \sum_{t=1}^{p} \beta_{4t} \Delta \mathrm{INFR}_{t-1} + \sum_{t=1}^{p} \beta_{5t} \Delta \mathrm{UUE}_{t-1} + \sum_{t=1}^{p} \beta_{6t} \Delta \mathrm{COVD}_{t-1} + \mu_{4t}$$

$$\Delta \mathrm{UUE}_t = \alpha_4 + \sum_{t=1}^{p} \beta_{1t} \Delta \mathrm{UUE}_{t-1} + \sum_{t=1}^{p} \beta_{2t} \Delta EG_{t-1} + \sum_{t=1}^{p} \beta_{3t} \Delta \mathrm{SER}_{t-1} + \sum_{t=1}^{p} \beta_{4t} \Delta \mathrm{INFR}_{t-1} + \sum_{t=1}^{p} \beta_{5t} \Delta RUE_{t-1} + \sum_{t=1}^{p} \beta_{6t} \Delta \mathrm{COVD}_{t-1} + \mu_{5t}$$

$$\Delta \mathrm{COVD}_t = \alpha_5 + \sum_{t=1}^{p} \beta_{1t} \Delta \mathrm{COVD}_{t-1} + \sum_{t=1}^{p} \beta_{2t} \Delta EG_{t-1} + \sum_{t=1}^{p} \beta_{3t} \Delta \mathrm{SER}_{t-1} + \sum_{t=1}^{p} \beta_{4t} \Delta \mathrm{INFR}_{t-1} + \sum_{t=1}^{p} \beta_{5t} \Delta RUE_{t-1} + \sum_{t=1}^{p} \beta_{6t} \Delta \mathrm{UUE}_{t-1} + \mu_{6t}$$

Where, Δ = first difference operator, p= optimal lag length and is the error term.

Where EG the economic growth of India, SER is the school enrolment ratio (higher), INFR is the inflation rate, RUE is the rural unemployment rate, UUE is the urban unemployment rate and COVD is the covid deaths.

Result and Discussion

India's economic growth

For examining the economic growth for any region or country GDP per capita is one of the important and efficient tool to infer. By examine the India's economic growth across year from 2000 to 2021 there is a gradual increase of GDP per capita but during the Covid-19 crisis (2019-2020) there is a steep fall in the GDP per capita of the country thus hindering economic growth.

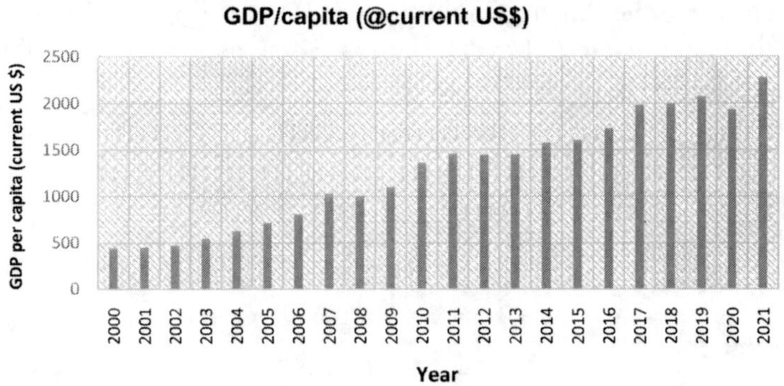

Fig. 1: GDP Per Capita of India (2000-2021).

Unemployment in India during Covid-19

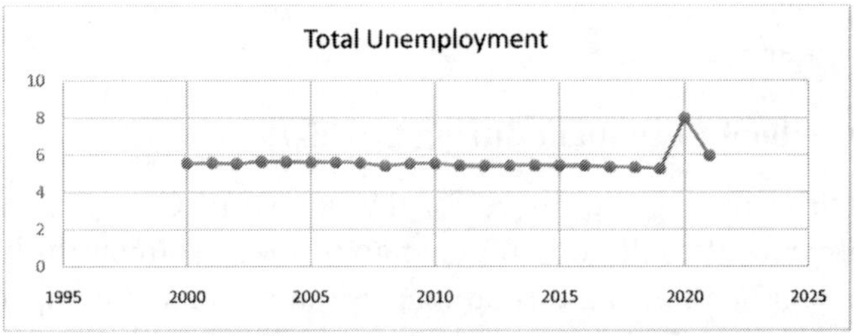

Fig. 2: Total Unemployment in India (2000-2020).

Source: ILO data, 2020.

By examining the past data of unemployment there is no trend, unemployment was there with a constant trend, but during the Covid-19 (2019-20) there is peak of unemployment in the overall India due to lockdown in the year 2020.

State-wise change in NSDP

For the state-wise analysis of Covid-19 effect the Net state domestic product was taken to represent the economic growth of each state by comparing it with during Covid and after Covid periods of three years viz., 2018-19, 2019-20 and 2020-21. The graph shows that there is increase in NSDP of each state during Covid except Jammu and Kashmir. This may be due to the exports which was done during Covid period. But the impact of Covid (2020-21) on each and every state is worse as indicated by the drastic decline in the NSDP.

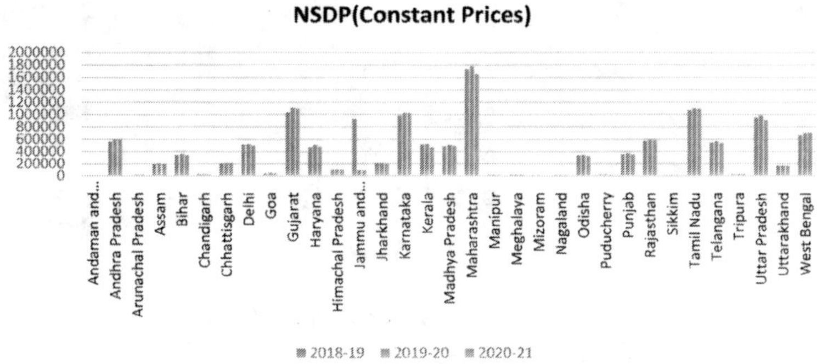

Fig. 3: Net State Domestic Product (2019-2020).

Source: RBI data, 2021.

High School Enrolment during Covid-19

On analysing the figure for the impact of Covid on school enrolment ratio, there is no change in the enrolment in higher education but also there is an increase in enrolment compared to previous year (2018-19) in almost each state except Andaman, Assam, Goa, Nagaland and Punjab. The reason for the increase in enrolment is due to the online classes by the every schools, both public and private and in Tamil Nadu the government led *Illam Thedi Kalvi* (Teacher going to student home for teaching) scheme for the rural students.

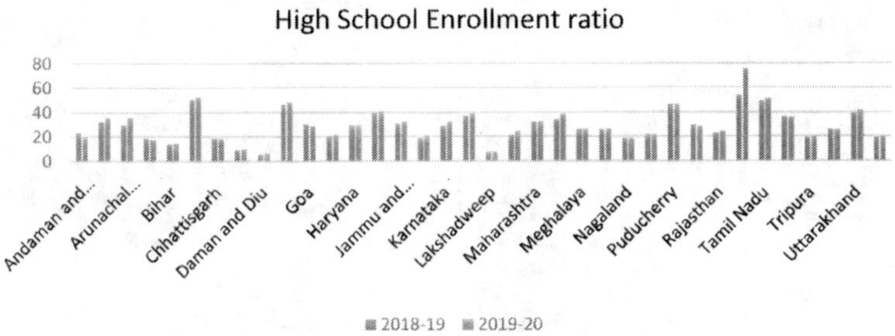

Fig. 4: High School Enrolment Ratio (2019-2020).

Source: RBI data, 2021.

Comparing Covid deaths across regions in India

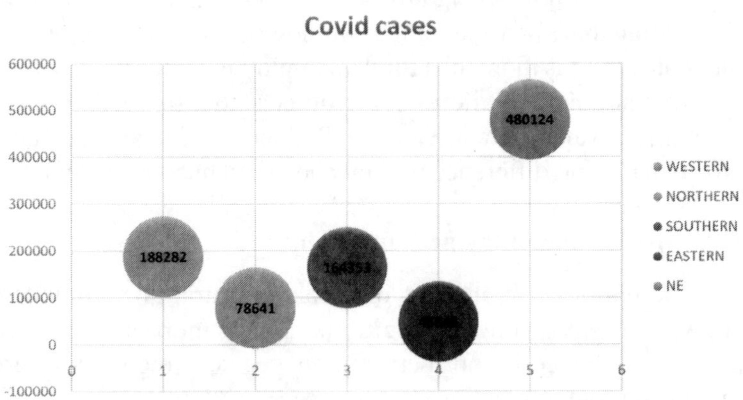

Fig. 5: Covid death across regions (2019-2020).

Source: MoHFW, 2022.

The figures explains in details that North-Eastern region experienced most deaths with 4,80,124 persons, followed by southern region with 1,64,353 persons, western region with 1,88,282 persons, northern region with 78,641 persons and finally eastern region with 48,848 persons.

State-wise comparison for rural and urban Unemployment during Covid-19

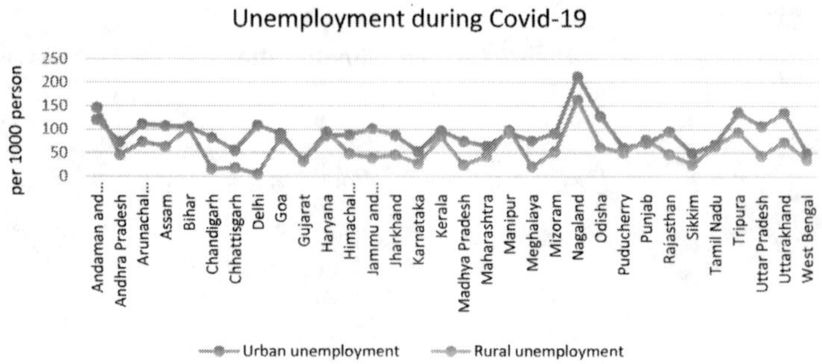

Fig. 6: State-wise rural and urban unemployment (2019-2020).

Source: RBI data, 2021.

Unemployment is the number of persons aged between 15 and 59 and not having any work in rural and urban regions. Comparing the urban and rural unemployment across states, urban unemployment is more than rural unemployment during Covid-19 except Andaman. While in some other states like Bihar, Goa, Gujarat, Haryana, Kerala, Manipur, Puducherry, Punjab and Tamil Nadu there is no difference in rural and urban unemployment.

Region-wise unemployment per thousand

The region-wise analysis of urban and rural unemployment shows that North-Eastern region is having more unemployed persons, followed by northern regions, eastern region, western region and southern region.

On comparing rural and urban unemployed persons across regions urban unemployment is more than rural unemployment in all regions during Covid (2019-20).

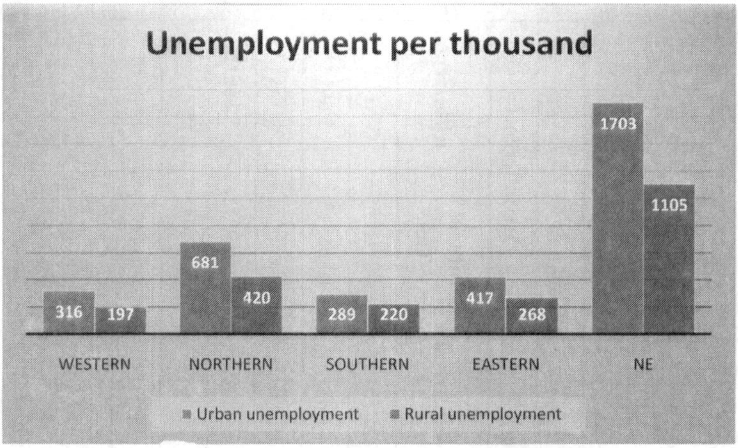

Fig. 7: Unemployed persons (2019-2020)

Source: RBI data, 2021

Empirical results

Unit root test

All states' cross-sectional data were checked for unit root by employing the augmented dickey-fuller (ADF) test and the Phillip-Perron (PP) test (PP test). The test results are established in Table 1. The results show that Economic growth and Covid deaths were stationary at levels across two tests, while the remaining variables were stationary at first difference form.

ARDL does not need unit root test for variables, but in this study it was used to conform the variables are integrated of order 1[I(1)] because this model is suitable if the variables are integrated of order 0 [I(0)], order 1 [I(1)]. The results also showed the suitability of ARDL model for cointegration for future research.

Granger Wald test for Causal relationship

The result for Granger causality Wald test is established in Table 2. The result shows that there are some unidirectional and no bidirectional relationship among the variables in our model.

Table 1. Result of Unit Root Test

	ADF test				PP test			
	Level		1st difference		Level		1st difference	
Variables	**without trend**	**with trend**	**without trend**	**with trend**	**without trend**	**with trend**	**without trend**	**with trend**
EG	-2.859*	-4.004	-9.284	-9.145	-6.144**	-5.761**	-9.094**	-10.124**
INFR	-0.161	-2.122	-6.710**	-6.813**	-2.138	-3.081**	-10.619**	-7.691**
SER	-1.510	-1.564	-9.201**	-7.207**	-0.518	-1.671	-8.125**	-8.734**
RUE	-2.913	-3.172	-6.915**	-4.274*	-3.064	-2.009	-6.819**	-5.156**
UUE	-4.965*	-3.277*	-9.621*	-9.606	-4.224**	-4.491**	-9.147**	-12.011**
COVD	-3.167*	-1.263*	9.251**	8.370**	4.401*	4.971*	8.481**	9.104**

There is unidirectional relationship between Inflation rate and economic growth which suggest that inflation rate will cause economic growth. There is no bidirectional causality observed among the variables during the study period.

There is unidirectional relationship between high school enrolment ratio and economic growth, rural unemployment and economic growth, urban unemployment rate and high school enrolment ratio, inflation rate and rural unemployment, high school enrolment ratio and rural unemployment. Covid deaths cause high school enrolment ratio, this is due to the online classes where parents are not willing to make admissions during the pandemic period.

Table 2. Granger Wald Test

Dependent variable	**Null hypothesis**	**Chi-Square**	**Probab-ility**	**Status**
EG	INFR does not cause EG	8.6669**	0.013	Reject
	SER does not cause EG	1.5608**	0.000	Reject
	RUE does not cause EG	8.275**	0.572	Accept
	UUE does not cause EG	3.2189	0.666	Accept
	COVD does not cause EG	2.5771	0.276	Accept
INFR	EG does not cause INFR	0.7555	0.685	Accept
	SER does not cause INFR	0.9623	0.618	Accept
	RUE does not cause INFR	0.9728	0.615	Accept
	UUE does not cause INFR	1.392	0.499	Accept
	COVD does not cause INFR	11.34**	0.003	Reject
SER	EG does not cause SER	0.4846	0.785	Accept
	INFR does not cause SER	2.1086	0.348	Accept
	RUE does not cause SER	0.9212	0.292	Accept
	UUE does not cause SER	1.9952**	0.013	Reject
	COVD does not cause SER	0.1127**	0.005	Reject

Dependent variable	Null hypothesis	Chi-Square	Probab-ility	Status
RUE	EG does not cause RUE	1.1349	0.567	Accept
	INFR does not cause RUE	9.1406**	0.010	Reject
	SER does not cause RUE	5.9733 *	0.050	Reject
	UUE does not cause RUE	1.5532	0.460	Accept
	COVD does not cause RUE	7.9313**	0.019	Reject
UUE	EG does not cause UUE	4.4709	0.285	Accept
	INFR does not cause UUE	1.2201	0.543	Accept
	SER does not cause UUE	0.6605**	0.019	Reject
	RUE does not cause UUE	0.1365	0.934	Accept
	COVD does not cause UUE	4.0446	0.132	Accept
COVD	EG does not cause COVD	0.6835	0.710	Accept
	INFR does not cause COVD	2.3493	0.309	Accept
	SER does not cause COVD	2.7017	0.259	Accept
	RUE does not cause COVD	1.4834	0.476	Accept
	UUE does not cause COVD	0.2916	0.864	Accept

Conclusion

This study was endeavoured to examine the socio-economic impression of Covid-19 on economic growth on Indian states using simple percentage, average and granger causality test. The percentage analysis for comparing economic growth and unemployment showed that during Covid-19 there is a negative growth in India as well all States and unemployment was its peak during 2019-2020. The region-by-region analysis revealed significant differences in unemployment, Covid-19 fatalities, and inflation rates.

Further, this study analysed the causal relationship among the variables and found that there are some unidirectional relationships among certain variables and no bidirectional relationship was observed.

Future scope of the study

This study used only the unit root test and the granger causality test, but this study may be extended to analyse the short-run and long-run relationship among the variables. And, also co-integration can be tested for the variables.

References

Barbate, V., Gade, R. N., & Raibagkar, S. S. (2021). Covid-19 and its impact on the Indian economy. *Vision*, *25*(1), 23-35.

Das, D., Kumar, K., & Patnaik, S. (2020). The impact of Covid-19 in Indian economy–an empirical study. *International Journal of Electrical engineering and technology*, *11*(3).

Debata, B., Patnaik, P., & Mishra, A. (2020). Covid-19 pandemic! It's impact on people, economy, and environment. *Journal of Public Affairs*, *20*(4), e2372.

Dev, S. M., & Sengupta, R. (2020). Covid-19: Impact on the Indian economy. *Indira Gandhi Institute of Development Research, Mumbai April*.

Kaushik, S., Kaushik, S., Sharma, Y., Kumar, R., & Yadav, J. P. (2020). The Indian perspective of Covid-19 outbreak. *Virusdisease*, *31*(2), 146-153.

Mehta, K., & Jha, S. S. (2020). Covid-19: a nightmare for the Indian economy. *UGC Care Journal*, *31*(20).

Rakshit, D., & Paul, A. (2020). Impact of Covid-19 on sectors of Indian economy and business survival strategies. *Available at SSRN 3620727*.

Ramakumar, R., & Kanitkar, T. (2021). Impact of Covid-19 pandemic on the Indian economy: A critical analysis. *Investigación económica*, *80*(315), 3-32.

Sidhu, G. S., Rai, J. S., Khaira, K. S., & Kaur, S. (2020). The Impact of Covid-19 pandemic on different sectors of the Indian Economy: A descriptive study. *International Journal of Economics and Financial Issues*, *10*(5), 113.

Singh, M. K., & Neog, Y. (2020). Contagion effect of Covid-19 outbreak: Another recipe for disaster on Indian economy. *Journal of Public Affairs*, *20*(4), e2171.

APPENDIX I

ANOVA

Source	Sum of squares(SS)	d.f.	Mean squares(MSS)	F ratio
Between region groups	22737183930	4	5.68E+09	1.200
Within region groups	1.08918E+11	23	4.74E+09	
Total	**1.32E+11**	**27**		

12

Covid-19, Inequality, and Marginalised Sections in India

V. Karthick

Introduction

The Covid-19 pandemic and its conundrum have not yet been over worldwide. Its enduring effect still affects society and economies. In the case of society, a few incidences in the USA and Europe have been reported that the Asian community were targeted due to racial prejudices during the recent pandemic period (Croucher et al. 2020; Rzymski and Nowicki, 2020). While developed countries continue to witness this kind of discrimination, developing countries have not been exempted, where the socioeconomically weakest sections of society have been the victims (Kipgen, 2020). In the case of the economy, irrespective of developed or developing or third-world countries, any form of disaster ultimately affects the poorest/marginalised sections of society, and Covid-19 is not exempted. The pandemic's aftermath affected most vulnerable communities across countries. The pandemics' resultant lockdown accelerates rising inequalities when a society has been divided based on prejudices and social segregation. Especially in India, the repercussion of the sudden, thoughtless, unscientific, and draconian lockdown affected 1.3 billion people, mainly historically enduring marginalised communities. Due to the pandemic, researchers have found that food and income primarily affect individuals' livelihoods. The pandemic may be inclusive, but their resultant curfews, quarantines, and lockdowns are not, as it excludes groups of individuals from access to socioeconomic resources such as employment, education, and health (Naskar, 2020; Deshpande and Ramachandran, 2020; Anand and Thambi, 2021). This pandemic-induced socioeconomic exclusion also led to job loss, income loss and food insecurity (Nair, 2020; Oxfam India, 2021; Ananthpur et al. 2022). Keeping this background, this paper explores the inequality due to the Covid-induced total lockdown impacting the socio-economically disadvantaged people across the Indian landscape.

This chapter is arranged as follows: Section II deals with whether the pandemic drives inequality, and Section III explores whether caste determines inequality. Section IV provides evidence of pandemic-induced inequality and its effect on marginalised sections. Section V discusses the issues before the conclusion in section VI.

Does the Pandemic Drive Inequality?

It appears that achieving the Sustainable Development Goals (SDG) by 2030 seems impossible because of persistent social and economic inequality due to the financial burden and natural disasters like Covid-19. For instance, Hoogeveens (2020) stated that achieving the SDG goals by the next decade does not seem to happen as we lag far behind due to a lack of money. On the other hand, the Goalkeepers report (2020), published by the Bill and Melinda Gates Foundation, indicated that nearly every indicator of the SDGs is off track for achieving them by 2030 due to the impact of Covid-19, the war in Ukraine and climate and food crises.

This is not the first time a pandemic has driven inequality, and studies have already mentioned it. For instance, Scheidel (2017) listed four types of catastrophic events, such as a pandemic, war, revolution, and state collapse found in human history, that have led to greater economic equality. The world has witnessed a massive

Covid-19 pandemic, which is still going through us. As a result, its resultant lockdown created widespread job losses and economic obstacles across income and occupational distribution. For instance, Deshpande and Ramachandran (2020) found that the virus is more likely to contract poorer and economically vulnerable populations. Also, they anticipated a higher risk of Covid-19 mortality among socially marginalised groups as they suffer from overlapping economic classes and social identities, like caste. In another article, Deshpande (2021) stated that the economic impact of Covid-19 has been much more demanding on those at the bottom of the caste ladder in India. Similarly, other studies also reported that, during the pandemic, members of Scheduled Caste (SC) and Scheduled Tribe (ST) suffered heavier job losses, income loss, etc. This paper further discusses these in detail in upcoming sections.

Whether Caste Determines Inequality?

The caste system in the country is the world's oldest social stratification that has been impacting education, employment, health, income, wealth, etc., for ages. The caste-lead inequality drives economic disparity. Because caste has always been a predictor of economic outcomes, it determines access to resources and opportunities. For instance, it determines income and expenditure (Desphande, 2000), capital and labour (Banerjee and Munshi, 2004), capital (Kijima, 2006), ownership of private enterprises (Thorat and Sabharwal, 2009a), education and occupation (Munshi and Rosenzweig, 2006), occupation and employment (Madheswaran and Attewell, 2007), children's nutritional status (Thorat and Sabharwal, 2009b), irrigation water (Anderson, 2011), access to – and – the amount of credit for agriculture (Karthick and Madheswaran, 2018, 2020a, 2020b; Madheswaran and Karthick, 2021 & 2022; Kumar, 2021) and maternal health and benefit (Mishra et al. 2021). Thus, it's been established that when caste influences everything in a country, it might also lead to inequality. In this context, this article put some evidence of inequality among marginalised sections due to Covid-19 in upcoming sections.

Pandemic-Induced Inequality and Its Effect on Marginalised Sections

Since the caste is an enclosed entity of class in India (Ambedkar, 2004), the pandemic accelerating inequality affects the society's marginalised sections primarily as their economic conditions are lower than other social strata. This article discusses the effect of inequalities persuaded during the time of Covid-19 in the country in the following subsections.

Agricultural Sector

As a whole economy, the pandemic-induced curfews lumped India's GDP significantly in the country's history by 24.4 percent between April and June 2020 (Dhingra and Ghatak, 2021). Agriculture faced severe income loss at the sector level as their marketable products were halted because of the travel and transport ban, and crops were left unharvested in the fields due to labour shortages. There were also shortages of agricultural inputs as shops were closed, and the next sowing got delayed due to shortages of labour and other inputs. Extreme poverty, hunger, debts, and farmer distress have already been prevalent among Indian agrarians. But these disadvantaged indicators aggravated during the pandemic (Jakhotiya, 2021). Out of 263 million workers employed in agriculture in India, 45 percent are cultivators, and the remaining 55 percent are agricultural labourers (Menon and Schmidt-Vogt, 2022). India constitutes mainly marginal and small farmers, i.e., 86 percent (Gulati and Juneja, 2019) and most of them belong to SCs and STs (Mahendradev, 2012). Since the marginalised sections of the farmers face severe agricultural discrimination regarding access to resources such as land, livestock and credit, the curfew and lockdown might affect them again. Thus, the effect of the pandemic would be much higher among them and lead to widespread inequality.

Unemployment and Job Loss

Both employment and job discrimination persist in the labour market due to the presence of caste in the country (Madheswaran and Attewell, 2007; Thorat and Attewell, 2007). The share of discrimination against SC/STs in 2019 was 39.3 percent (Kundu and Khan, 2022). Covid-19 and its effect worsened this situation. The initial months of the severe lockdown witnessed a sharp rise in unemployment. An estimate shows that 119 million workers of SCs have lost their jobs within the two weeks of the lockdown (Naskar, 2020). If those households' average family size is five as per the 2011 census, then they could be facing a severe livelihood crisis. Deshpande and Ramachandran (2020) examined that the proportion of employed SC and ST categories dropped from 44 percent to 24 percent and 40 percent to 26 percent, respectively,between December 2019 and April 2020. The percentage of falls for SCs and STs was 20 and 14 percentage points, respectively, and it is almost three and two times as large as the General category, as their fall was seven percentage points. This result witnesses a greater fall in employment for both SCs and STs than for the General category. In the case of job loss, Deshpande and Ramachandran (2020) reported that the rate of job losses among SC/STs was higher than the General category immediately after the lockdown. Since the latter have more access to education and more secure jobs, they are less vulnerable to economic shocks due to the pandemic and lockdowns. But the job loss cases among SC/ST households have been higher as many workers in this category are engaged in casual work. Their earnings are also meagre as they receive 55 percent less than the General caste receives (Anand and Thambi, 2021).

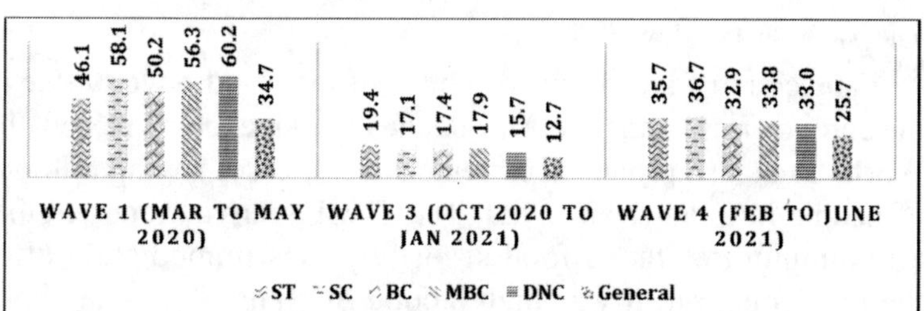

Fig. 1. Share of Job Losses among Marginalised Caste Groups in Tamil Nadu (%).

Source: Ananthpur et al. (2022).

Corroborating with the findings of Deshpande and Ramachandran (2020), a study based on the Tamil Nadu Covid Pulse Survey (TNCPS[1]), Ananthpur et al. (2022) also found the same results. The survey observed a higher percentage of job loss among Denotified Communities (DNC) and SC households (60% and 58%, respectively) during wave 1. But at the end of wave 4, the scenario changed. The job loss share was high among SCs (36.7%) and STs (35.7%). It has been reported that the caste groups association with vulnerable work in sanitation and construction worsened the situation during that time. Throughout the lockdown, a reduction in job loss has been reported among all social groups compared to the first wave. But during wave 4, the job losses faced by ST and SC were higher than the other social groups across (Fig. 1).

Food Insecurity

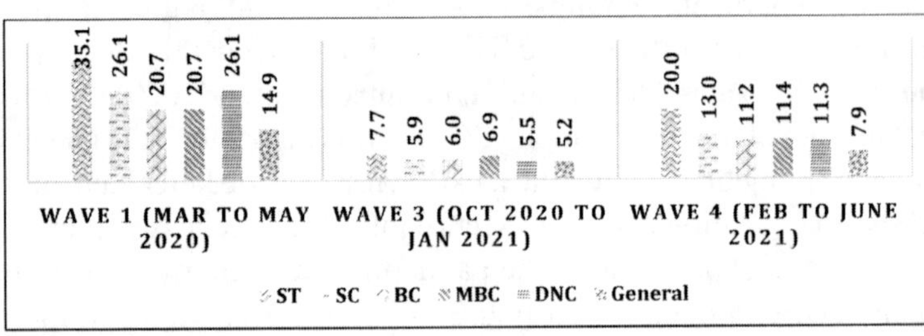

Fig. 2. Food Security Issues across TNCPS Waves among Social Groups (%).

Source: Ananthpur et al. 2022.

The country faced severe food insecurity and poverty due to the corona virus pandemic. During the initial lockdown period, the World Bank also projected extreme poverty cases among millions of people in India (Nair, 2020).The TNCPS survey reported that all communities faced food security issues immediately after the pandemic, but it was high among SCs and STs (Ananthpur et al.2022). During the first wave, increased food insecurity was observed among STs and SCs (Fig. 2). They faced a 35 percent and 26 percent food shortage compared with other categories. Whereas in Round 4, with the explosion of the second wave of Covid-19, households from the ST and SC categories remained top in food insecurity issues. Their share of food insecurity was 20 percent and 13 percent, respectively, than other categories. The results showed that ST and SC categories remain the most vulnerable and affected communities.

Health Inequality and Physical Distancing

Around 32 percent of the ST/SC households have single-room dwellings, and 14 percent live in urban slums (Anand and Thambi, 2021). Since there is not enough room to follow the Covid-19 protocols to maintain physical distance, the chances of SC/ST household members getting infected are higher than the General category. Another study from Oxfam India (2021) came with shocking results of health inequalities during Covid-19. This was a primary survey conducted in seven states consisting of 768 respondents on with Covid-19 or having recovered from Covid-19. The study showed that the pandemic had exacerbated inequalities among marginalised groups already facing disproportionate health treatment due to growing socioeconomic disparities. Further, the report also indicates that SC/STs who faced difficulties in accessing non-Covid medical facilities are over 50 percent compared to 18.2 percent in the General category. As Taneja, affiliated with Oxfam India's Inequality, Health & Education report,said, *"the current health status of the country is a testament to the unfulfilled dream of 'Health for All'. The right to the highest attainable health is far from being realised"*.

Education and Access to Technologies

Forages, education has been denied to a particular section of the population. It is closely related to mobility and returns in the labour market (Anand and Thambi, 2021). The study also found that ST/SC's never-enrolled proportions were significantly higher, even though the dropout ratio also was higher among those who enrolled. While engagement in economic activities was reported as the primary reason for dropout, denying education was equally important. Further, the disadvantaged social and economic groups were in a new crisis in the education sector during the pandemic and its subsequent recession. Because of the lockdowns, education moved online and enrolled members of households have not attended classes due to a lack of access to digital devices like computers and an internet connection. Before the pandemic, percentages of households with at least one member enrolled in an educational institution that could access a computer and access the internet were about 12 percent and 27 percent,respectively,in 2017-2018 (Fig. 3). Among them, ST/SC household's share was significantly lower than other categories in these respects. For instance, Figure 3 shows that the percentage of accessing a computer at home for ST and SC households are 5.7 percent and 5.9 percent, respectively,which are lower than other categories.

Similarly, the share of access to the internet for these categories is 13.1 percent and 17.8 percent, respectively. These households are already facing severe job loss and socioeconomic disadvantage, and the pandemic-induced lockdown has also affected their education. Because of their lower income and job loss, increasing dropouts in these households were more likely during the lockdown. As a result, the exclusion from accessing online education and thereby completing the education to enter the job market is perilous.

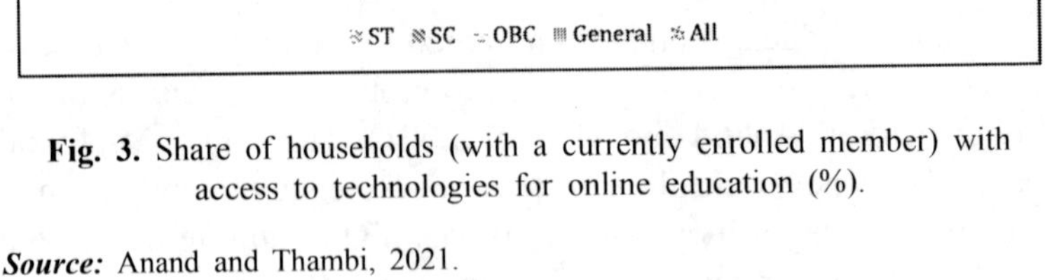

Fig. 3. Share of households (with a currently enrolled member) with access to technologies for online education (%).

Source: Anand and Thambi, 2021.

Deshpande and Ramachandran (2020) find that fewer savings and investments among the SCs also led to differential access to information technology in terms of access to online education. For instance, SCs with bank savings are just 49 percent compared to 62 percent of General category households. This would reflect disparities in the ability to invest in technology among disadvantaged communities if the pandemic forces schools to close for a substantial period.

Conclusion

The results presented in the above sections reveal that most Indians faced substantial inequalities by multiple axes before and after the pandemic. The marginalised sections, especially SC/ST, already live in poverty, face an income crisis, have lower access to health, education, and other resources, and are also denied possessing economic resources. Additionally, the Covid pandemic and its resultant and stringent total lockdown accelerate inequality among them. SC/STs are deprived of landholdings, but the rising unemployment among them also increases income loss and leads to poverty. Thus, their health would be at stake. Since they have less investment and savings, their children's education is also threatened. The studies mentioned above reported that the extent of distress due to the lockdowns and the uneven burden of the pandemic exposed and deepened the crisis of inequalities,further worsening the situation among marginalised sections.

Due to this, achieving Goal No.10 of SDG, i.e., reducing inequality by 2030, seems impossible. Investing in education, employment, livelihood, and health is essential as that might close the gap between social groups. Similarly, targets 10.2 and 10.3 of Goal 10 must be taken very seriously. Target 10.2 aims to empower and promote the social, economic, and political inclusion of all, regardless of age, sex, disability, race, ethnicity, origin, religion or economic or other status, to reduce inequality. On the other hand, Target 10.3 intended to ensure equal opportunity and reduce inequalities of outcome, including eliminating discriminatory laws, policies and practices and promoting appropriate legislation, policies, and action in this regard (Target 10.3).Therefore, there should be stringent laws against discriminatory practices; otherwise, achieving social, economic, or any other equality is impossible by 2030 or even after that.

Note

1. The survey collected the information related to Covid and its effect on household in Tamil Nadu using CATI (Computerised Assisted-Telephonic Interview) method during April 2020 to June 2021 with four-time interval.

References

Ambedkar, B R (2004). Castes in India: Their Mechanism, Genesis and Development. Readings in Indian Government and Politics Class, Caste, Gender, 131-53.

Anand, I., and Anjana T. (2021). The Crisis of Extreme Inequality in India. *The Indian Journal of Labour Economics* (2021) 64:663–683. https://doi.org/10.1007/s41027-021-00335-9

Ananthpur, K., Jafar, K., Sridharan, G., and Venkatachalam, L. (2022). Impact of The Covid-19 Waves on Tamil Nadu's Economy and Society. *MIDS Working Paper No. 241*, Madras Institute of Development Studies.

Anderson, S. (2011). Caste as an impediment to trade. *American Economic Journal:Applied Economics,3*(1), 239–263.

Banerjee, A. V., and Munshi, K. (2004). How efficiently is capital allocated? Evidence from the knitted garment industry in Tirupur. *TheReview of Economic Studies,71*(1), 19–42.

Croucher, S., Rahmani, D., Nguyen, T. (2020). Prejudice toward Asian Americans in the Covid-19 pandemic: the effects of social media use in the United States. Front Commun 5:39.

Deshpande, A. (2000). Does Caste Still Define Disparity? A look at Inequality in Kerala, India. *The American Economic Review*, 90,322–325.

Deshpande, A. (2021). How India's Caste Inequality Has Persisted—and Deepened in the Pandemic, *Current History,* 120 (825), 127–132

Deshpande, A., and Ramachandran, R. (2020). Differential impact of Covid-19 and the lockdown, *The Hindu* (retrieved on 22 August 2020).

Dhingra, S., and Ghatak, M. (2021). The Pandemic in Data: How Covid-19 Has Devastated India's Economy, *Scroll.in*. Available online: https://scroll.in/article/999275/the-pandemic-in-data-how-Covid-19-has-devasted-indias-economy (accessed on 30 January 2023).

Hoogeveens, H. (2020). UN will not be able to achieve sustainable development goals by 2030, *Mint*, (retrieved on 20 August 2020).

Jakhotiya, G. (2021). Impact of Covid-19 on Indian economy and the road ahead. MPRA 2020. MPRA Paper No. 101456. Available online: https://mpra.ub.uni-muenchen.de/101456/ (accessed on 6 August 2021).

Karthick, V., and Madheswaran, S. (2018). Access to formal credit in the Indian agriculture: Does caste matter? *Journal of Social Inclusion Studies,4*(2), 169–195. https://doi.org/10.1177/2394481118814064

Karthick, V., and Madheswaran, S. (2020a). *Whether Caste Impedes Access to Formal Agricultural Credit in India? Evidence from NSSO Unit Level Data*, ISEC Working Paper 478, Institute for Social and Economic Change, Bengaluru.

Karthick, V., and Madheswaran, S. (2020b). *Determinants of Agricultural Credit in Rural India by Social Group*, ISEC Working Paper Series 490, Institute for Social and Economic Change, Bengaluru.

Kijima, Y. (2006). Caste and tribe inequality: Evidence from India, 1983 to 1999. *Economic Development and Cultural Change,54*(2), 369–404.

Kipgen, N. (2020). Covid-19 pandemic and racism in the United States and India. *Econ Political Wkly*. 55(23), 21–6.

Kumar, A., Das, R., Aditya, K. S., Bathla, S., & Jha, G. K. (2021). Examining institutional credit access among agricultural households in Eastern India: Trends, patterns and determinants. *Agricultural Finance Review, 81*(2), 250–264. https://doi.org/10.1108/AFR-04-2020-0054

Madheswaran, S., and Attewell, P. (2007). Caste discrimination in the Indian labour market: evidence from the national sample survey. *Economic and Political Weekly,42*(41), 4146–4153.

Madheswaran, S., and Karthick, V. (2021). "Access to Institutional Credit by Caste among Agricultural Households in India: A Decomposition Analysis,"*Aarthika Charche: FPI Journal of Economics and Governance*, 6(1),77-92. https://fpibengaluru.karnataka.gov.in/storage/pdf-files/Aarthika percent20charche/AarthikaCharche-Vol6-No1-2021.pdf

Madheswaran, S., and Karthick, V. (2022). "Caste Discrimination in Agricultural Credit Markets" published a chapter in India Discrimination Report 2022, edited by Prof. Amitabh Kundu, published by Oxfam India. pp.48-53. https://www.oxfamindia.org/knowledgehub/workingpaper/india-discrimination-report-2022

Menon, A., and Schmidt-Vogt, D. (2022). Effects of the Covid-19 pandemic on farmers and their responses: A study of three farming systems in Kerala, South India. *Land*, 11(1), 144.

Mishra, P. S., Veerapandian, K., & Choudhary, P. K. (2021). Impact of socioeconomic inequity in access to maternal health benefits in India: Evidence from Janani Suraksha Yojana using NFHS data. *Plos one*, *16*(3), e0247935.

Munshi, K., and Rosenzweig, M. (2006). Traditional institutions meet the modern world: Caste, gender, and schooling choice in a globalising economy. *The American Economic Review, 96*(4), 1225–1252.

Nair, R. (2020). India could see millions pushed into extreme poverty due to pandemic: World Bank analysis. *The Print* [Published online, 12 June 2020, Accessed 10 November 2022] https://theprint.in/economy/india-could-see-millions-pushed-intoextreme-poverty-due-to-pandemic-world-bank-analysis/440298/

Naskar, S. (2020). Social Distancing, Caste Discrimination and Vulnerability amidst Covid-19 Pandemic.

Rzymski P., and Nowicki M. (2020). Covid-19-related prejudice toward Asian medical students: a consequence of SARS-CoV-2 fears in Poland. *J Infect Public Health*, 13(6), 873–6.

Scheidel, W. (2017). *The Great Leveler: Violence and the History of Inequality from the Stone Age to the Twenty-First Century*, the Princeton University Press.

Thorat, S. and Sabharwal, N.S. (2009a). Caste and ownership of private enterprises. *Economic and Political Weekly, 44*(23), 13-16.

Thorat, S. and Sabharwal, N.S. (2009b). Discrimination and children's nutritional status in India. *Institute of Development Studies Bulletin, 40*(4), 25-29.

Thorat, S., and Attewell, P. (2007). The legacy of social exclusion: A correspondence study of job discrimination in India. *Economic and Political Weekly,42*(41), 4141–4145.

13

Climate Change and Rice Productivity in Tamil Nadu: Advocating Possible Adaptive Capacity

S. Smirthi Rajya, Aswathy Ann Thomas and
Merlin Juliet Arul Thangam

Introduction

Climate change is one of the burning issues in the recent times, around the world and its impact is widening day by day. Half a degree increase in the level of warming may not sound much but its effects might create dramatic difference, like endangering millions of populations through heat waves, water scarcity, coastal floods, droughts etc., The United Nations has come up with Sustainable development Goals for the future of the society to face these serious issues around the world. And ***this paper is closely correlated with the principles of 13th Sustainable Development Goal that is Climate Action***. The main aim of this goal is to fight against climate change and its impacts on the environment around us. This paper tries to create at least a minimal awareness on the current situation of the agricultural sector due to the impact of climate change. India being the second largest populated country, with more than 50 percent of its population still focusing on agriculture for its employment, contributing almost 17 to 18 percent towards the country's GDP. It is indeed entitled to safeguard itself from all these destructive results of climate change. That too, cater to the huge population of the country is still a great challenge and climate change is just adding fuel to the fire. Being cognizant of the situation, this study attempts to analyse the influence of the climate variables (Temperature and Rainfall) on the rice productivity of Tamil Nadu. The reason for choosing one single state and its staple food is to conduct the study in the regional/local scales to get a more detailed picture about the required study, because of the large regional differences in the surface temperature.

Tamil Nadu is the 12th largest state with agricultural sector contributing almost 16.88 percent towards the economic development of the state. Over 70 percent of the state's population is dependent on agriculture for its employment, yet this sector contribution is only 13 percent towards the state economy. There are numerous reasons for this inappropriate economic contribution in the agricultural sector and climate change is one among them. Tamil Nadu is a major rice growing state in India enriched with all favourable tropical climatic condition for its growth. The agricultural zone of Tamil Nadu is classified into seven agro-climatic zones depending on the rainfall distribution, soil fertility, irrigation patterns, crop cultivation and other physical, ecological and geographical factors (K & R, Status Paper on Rice in Tamil Nadu). The total cultivable area of Tamil Nadu is 8.16 M ha, out of which 2.80 M ha is irrigated land and the remaining 5.36 M ha is rain fed land (K & R, Status Paper on Rice in Tamil Nadu). The state receives an annual average rainfall of about 937.5 mm, where the majority of the rainfall is received by the state during the North-east monsoon (October - December) (K & R, Status Paper on Rice in Tamil Nadu). The state cultivates many crops like paddy (Rice), Millets, Pulses, Oil seeds, Cotton, Sugarcane and Coconut but still we are only focusing on the productivity of rice alone in this study, because of its importance in the state and for the purpose of ease in understanding the effect of climate change.

A considerable amount of literature was read from the same background dealing with rice productivity and climate change and it was analysed that there is indeed a significant level of impact on rice productivity due to the carbon footprint in the recent years and yet we are not provided with sufficient evidence to prove it strongly that climate change is a major damage causing agent. This situation calls for the requirement of possible adaptive measures that has to be taken to reverse the situation. Thus, this becomes the essence of this study.

A bird's eye view on Tamil Nadu's Paddy Cultivation

Paddy is one of the prime crops that is cultivated in almost all the districts of the State in a unique three-season pattern viz., Kar, Kuruvai, Sornavari (April to July), Samba/ Thaladi, Pishanam (August to November) and Navarai, Kodai (December to March) (K & R, Status Paper on Rice in Tamil Nadu). There are almost 8 different varieties of rice grown in the state of TamilNadu during the different seasonal conditions which is shown in the table. The duration for the cultivation of rice generally ranges between 110-115 days, 125-135 and 140-160 days (K & R, Status Paper on Rice in Tamil Nadu).Thus, it is evident that the cultivation of rice mainly depends on rainfall. Further in this state the majority of the cultivation depends on the rain fed land and along with the onset of climate change it is indeed necessary to observe its production pattern with the recent fluctuations in rainfall patterns and changes in the temperature.

Name of the Rice	Growing Season
Sornavari	March-April to June-July
Kar	April-May to July-August
Kuruvai	May-June to August
Early samba	July-August to Jan-Feb
Samba	Sept-Oct to Jan-Feb
Late samba	Sept.-Oct to Jan-Feb
Thaladi / Pishanam	Sept-Oct to Feb-March
Navarai	Nov-Dec to Feb-March

Source: Status Paper on Rice in Tamil Nadu, Centre for Plant breeding and Genetics, Tamil Nadu Agriculture University, Coimbatore.

Objectives

The following are the objectives of the study: This paper tries to capture the effect of climate change using the basic variables like temperature and rainfall on the productivity of rice, which is one of the staple foods of Tamil Nadu. The period of the study is between 2005-2021.

To analyse the trends on the productivity of rice in TamilNadu using the variables like temperature and rainfall from the period of 2005-2021.

To provide future scope for adaptability and mitigation that best suits the condition of the state of Tamil Nadu, after analysing the existing strategies for adaptation and mitigation of climate change activities, with respect to rice production.

Literature Review

The motivation for this study is gained by having a vast idea from other scholarly works with the same outlook. The research work by Saravana Kumar (2014), tries to capture the difference in productivity of almost all major crops of Tamil Nadu with respect to Climate variables and the results obtained that the productivity of rice and sorghum are under risk in the near future and there is a requirement for adaptation measures to be taken with no delay. The research work by Md. Abiar Rahman et al. (2017) tries to investigate the influence of temperature and rainfall on the productivity of rice in Bangladesh with various ecosystem like dry, terrace and coastal and the results were obtained using Standardised precipitation index, Diurnal temperature range and Rice productivity Index. The results obtained states that the rice productivity is under risk and stresses the importance of mitigation measures to tackle the adverse situation. The scholarly work of J. Srinivasan (2019) intends to analyse the results of the existence of climate change in India and provides the fact that the changes in level of temperature, rainfall patterns, glaciers, sea level rise, drought etc are all the major factors that are affected because of climate change. And we are in immediate need of adaptation and mitigation measures to reverse the situation to at least to the least level possible. Thus, with these scanning we can understand the necessity of more research in this area for fastest revival of the scenario.

Data Collection and Methodology

The idea behind this study is to observe that the prime variables (Temperature and Rainfall) of climate change has a significant effect on the rice productivity, in the state of TamilNadu. To examine the same, we have determined to establish a trend analysis between the rice productivity (Metric tonnes) and the temperature (Fahrenheit) and Rainfall (Millimetres') that the state receives from the period of 2005 to 2021. The rice productivity data for the above trend analysis is obtained from the Director of Agriculture, Department of Agriculture, Chennai. The temperature data for the same period was obtained from the website wunderground.com and the rainfall data for the same period was obtained from the Indian Meteorological department, Chennai.

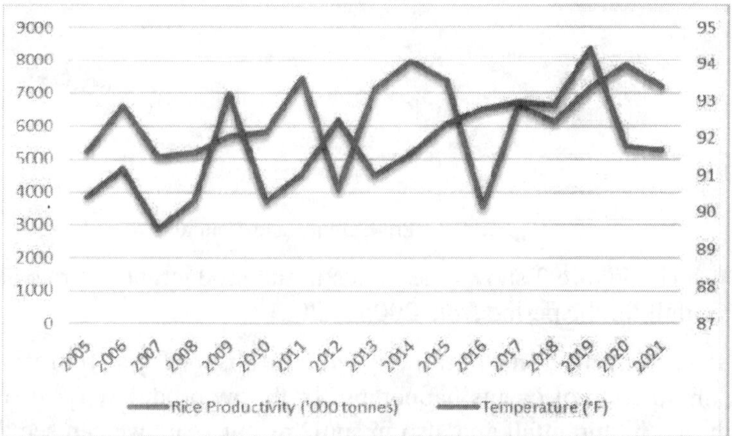

Fig. 1: Rice Productivity and Temperature.

From the above figure 1 we can identify that whichever year records the high level of temperature for the corresponding year there is a decline in the productivity of rice. This shows that there is negative relationship that exists between these two variables and provides the critical information that temperature plays a crucial role in rice productivity, along with other abiotic and biotic factors. The recent report states that the increase in temperature from the last 150 years, primarily accounts on 40 percent increase in carbon dioxide (Srinivasan, 2019). This rings the alarm that even the minute rise in the temperature due to climate change should not be left unattended. The preference for adaptive and mitigation measures are to be implemented without any delay. Further these measures have to be implemented by each and every responsible individual along with the Government.

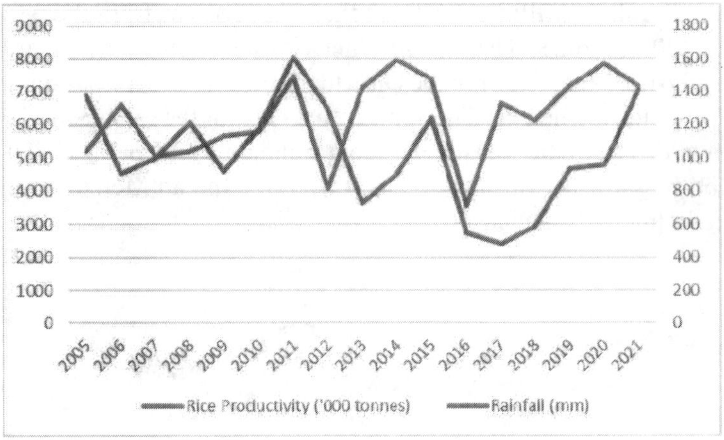

Fig. 2: Rice Productivity and Rainfall.

The Figure 2 shows the trends in the productivity of rice and rainfall for the period from 2005 to 2021.

From the trend obtained,. we can observe that there is a combination of events happening. Both low productivity of rice during high rainfall and also in some recent years we can see the productivity of rice increasing during low rainfall. Thus, we can conclude that both too little rainfall and too heavy rainfall will damage the productivity of the rice crops. Thus, it is always essential for the government to properly channelise all the excess rainfall in the state to be used during the summer seasons, which is still a challenging task. Tamil Nadu receives the majority of the rainfall during the North-east Monsoon, which is different from all the others state in India. On an average the state receives about 925 mm of rainfall per year and the major cultivable land in the state depends on rainfall rather than the irrigation facilities. Thus, with these two trends we can infer that for a flourishing productivity of rice or any other crop we need to focus on the temperature and rainfall patterns. To stabilise these two situations for a sustainable future, the only solution that we can rely on is to focus on adaptation and mitigation processes. For this we need to know the already existing policies of the states so that we can suggest more modifications for a progressive recovery in the future.

State's Plan towards Climate Resilience

According to the sixth Assessment Report of IPCC, 29 districts in TamilNadu are identified as vulnerable to climate change. Rice is widely cultivated in all these districts of TamilNadu.

Thus, it is inevitable to take proper measures in improving the production of these agricultural products. The Government of Tamil Nadu has also provided its attention in managing climate change through its policies and schemes. We shall discuss a few such schemes and policies with special focus on rice productivity. (Government of Tamil Nadu, Department of Environment and Climate Change, n.d.)

In accordance with the National Action Plan on Climate Change (NAPCC) initiated by the Government of India, in the year 2008, the state Government of Tamil Nadu has enforced the **TamilNadu State Action Plan on Climate Change** (TNSAPCC) (Government of Tamil Nadu, Department of Environment and Climate Change, n.d.) in the year 2015 to address the climate change issues and to promote developing goals.

The TNSAPCC had identified seven important regionally vulnerable sectors namely ***Sustainable Agriculture***, Water Resources, Forest and Biodiversity, Coastal Area Management, Sustainable Habitat, Energy Efficiency, Renewable Energy and

Solar Mission, Knowledge Management to work with to reduce the effect of climate change. In 2020, The Government of Tamil Nadu upgraded the **TNSAPCC - 2.0** with 37 activities as long-term implementation plan along with the already existing 7 action plans.

To increase the productivity of all crops the state has implemented **National Agricultural Development Programme and National Food Security Mission Schemes** in the year 2022-2023.

The Government of Tamil Nadu has announced Three big ticket initiatives viz: **(i) Tamil Nadu Climate change Mission (ii) Tamil Nadu Wetlands mission (iii) Green Tamil Nadu Mission** (The New Indian Express, 2021), to focus on climate change adaptation and mitigation activities.

Free tree samplings are being distributed to farmers to increase the green cover of the state in the name of **Tamil Nadu Mission for Sustainable Green cover in Farm Lands**, which is funded by both the central and the state.

The Government of Tamil Nadu has encouraged its farmers to switch to **Organic farming**, by providing **Green Manure Seeds**. This will help in improving the availability of organic products and also helps in maintaining the soil fertility as well.

The Government of Tamil Nadu has implemented some special schemes to increase the productivity of rice in order to encourage the cultivators.

Thiru. C. Narayanasamy Naidu Paddy Productivity Award: This award is to recognise the farmers with a cash prize, who obtain highest productivity of rice using the SRI (The System of Rice Intensification) method of cultivation. Even though this is one of the labour-intensive methods of cultivation, it is encouraged as it requires less water. (Government Of Tamil Nadu, Department of Environment and Climate Change, n.d.)

Bharat Ratna Dr. M.G.R. Traditional Rice Variety Conservator Award: This award is to acknowledge the framers who still conserve and cultivate the traditional Paddy varieties.

This will motivate them to preserve the age-old traditional rice varieties to the generations to come. (Government of Tamil Nadu, Department of Environment and Climate Change, n.d.)

These are some of the legitimate policies and schemes that the government of Tamil Nadu is currently following to reduce the ill effects of Climate change to keep the production of rice in par with its demand.

Current Barriers and Future Perspectives:

The latest Anna University study says that the average day temperature to increase upto three percent by 2100, which is ten times faster in the warming level than 1000 years before. The National centre for coastal research has identified that 41 percent of the state's coastal areas are eroding. The Tamil Nadu state Action on Climate Change states that rainfall will decrease upto nine percent by 2100.

Thus, it is high time for us to realise that the evidence of Climate Change has spread all around us and the future of the world depends on the efficient enforcement of the adaptation and mitigation measures. These measures have to be implemented both globally and regionally, the global objective should focus on the wider effects like controlling the CO_2 (Carbon dioxide) and other greenhouse gas emission and the regional mission should focus more on controlling the various natural and man made pollution in their respective territory to reduce the jolt of climate change.

Conclusion

Adaptation along with the proper mitigation efforts is considered as the paramount response in resisting against climate change. In this aspect the state of Tamil Nadu has numerous policies and strategies to improve the condition of agriculture, through many planned short- and long-term visionary schemes. Yet the state faces many challenges in achieving the same because of the major fact that there is still lack of knowledge and support among the common people with respect to the ramifications of climate change. Many research organizations state that the competency of any adaptation or mitigation measure purely depends on the potential or the extent to which the affected community or the region is ready to cope with the adversity of the same. Thus, keeping this in mind, it is required to educate the common people about the devastating outcome of climate change and the urgent need to focus on sustainable lifestyle to have a future.

One such great step forward is the initiation of a separate Ministry for Environment by the current Government will be the beginning of the success story. The state of Tamil Nadu should further concentrate on the cultivation in the rain fed areas with special attention towards the 29 vulnerable districts towards climate change. Wherein the state has to incorporate **Climate Smart Agriculture,** which is the method of including both the traditional and newly developed technology in increasing the productivity of the agricultural crops. Such as, it is preferable to seek the help of technology in finding new seed varieties that are less sensitive to rainfall and temperature, thereby getting adapted to climate change. Also,it is emphasised to shift from traditional sowing patterns of the seeds as the old method does not provide expected result with the current changes in the factors supporting agricultural growth.

In addition to all these, the State also has to keep an eye on the **effect of environmental migration** that might happen if we are not implementing these adaptation and mitigation measures promptly. It might lead to local or global conflict, increase the gap between rich and poor and might lead to overall disintegration in the state or in a nation.

Thus, with these inputs, this study attempts to throw some light on the importance of timely implementation of adaptive and mitigation measures in battling against the consequences of Climate change towards the agriculture productivity, which is still the backbone of the economy.

References

Dubash, N. K. (2019). An Introduction to India's evolving Climate change - Debate: From diplomatic Insulation to policy integration . In N. K. Dubash, *India in a Warming world - Integrating Climate change and Development* (pp. 1-30). New Delhi: Oxford University Press.

Government of Tamil Nadu, Department of Environment and Climate Change. (n.d.). Retrieved from www.environment.tn.gov.in.

K, T., & R, K. *Status Paper on Rice in Tamil Nadu*. Hyderabad: Rice Knowledge Management Portal.

Md. Abiar Rahman, S. K. (2017). Impacts of temperature and rainfall variation on rice productivity in major ecosystems of Bangladesh. . *Agriculture and Food Security*, 6–10.

Medhavi Sandhani, A. P. (2020). Impact of Climate change on Economic Growth: A case study on India . Madras School of Economics.

Pulak Guhathakurta, U.K. (ESSO/IMD/HS/Rainfall Variability/ 24(2020)/48). *Observed Rainfall Variability and Changes Over Tamil Nadu State*. GOVERNMENT OF INDIA, Ministry of Earth Sciences, INDIA METEOROLOGICAL DEPARTMENT .

Saravanakumar, V. (2014). Impact of Climate change on yield of major food crops in Tamil Nadu, India . *Working Paper - South Asian Network for Development and Environmental Economics (SANDEE)*, No. 91-14, p. 25.

Shyamasree Dasgupta, A. B. (2019-2020). *Climate vulnerability and risk assessment at the national level using a common framework*. Department of Science & Technology.

Srinivasan, J. (2019). Impact of Climate change on India. In N. K. Dubash, *India is a warming world—Integrating Climate change and Development* (pp. 31-44). New Delhi: Oxford UNiversity Press.

Xin Xuan, B. L. (2021). Climate Change and Adaptive Management: Case Study in Agriculture, Forestry and Pastoral Areas. (pp. 10, 832). MDPI.

14

Plastic Waste Economics and an Alternative Solution to this Plastic Menace

Megha Mahajan, T.G. Nithya and *M. Kamaraj*

Introduction

Environmental pollution is increasing day by day due to many factors, and among them, one of the major causes of this nature disturbance is the overuse of plastics. Our environment contains micro plastics and other contaminants made up of plastic, which are a great threat to all living organisms living on this mother earth (Dohler et al. 2022; Moshood et al. 2022). Plastic pollution not only poses a great threat to the oceans, environment, and living organisms, but it is also entering the food chain in the form of micro plastics. These micro plastics are everywhere: in the air, water, and soil, and there have been reports and studies suggesting that these micro plastic traces are also found in human blood (Cox et al. 2019). So, in this world of urbanization, there is a need to solve this problem and find an alternative route to this environmental issue. This issue might be resolved by using biodegradable plastics, which focus on creating a more environmentally friendly, sustainable world. Biodegradable plastics, which have properties similar to commercially available conventional plastics, are also responsible for lowering carbon dioxide levels in the environment. To reduce waste management, there is an increasing demand for affordable, environmentally friendly materials (Oktavilia et al. 2020).

Excessive production of plastic, which was once thought to offer substantial benefits, has altered due to improperly handled plastic waste, turning into a serious hazard to the biosphere (Sakti et al. 2021). In reality, more than 700 species of marine life have been put at risk by mismanaged plastic trash in the last 20 years, and this number is predicted to keep climbing as long as plastic waste is poured into the ocean, according to a Gall and Thompson (2015) analysis. Marine ecosystems are natural resources for the more than 3.5 billion people on Earth who depend on them for their livelihood (Thompson et al. 2004). Studies show that between 1.5 million and 2.4 million metric tonnes of plastic waste are dumped into the ocean each year (Lebreton et al. 2017).

Plastics production statistics

Plastics today provide society with a number of advantages. They not only help feed the world in a safe manner but also help create the world due to properties such as easy availability, lower cost, high tensile strength, the ability to be moulded into any shape, and so on. Even better, they can keep us alive. Without a doubt, plastics are important materials for innovation, and our industry must also make sure that plastics are sustainable and have a beneficial influence on both people and the environment because they are vital materials for society (Plastics Facts, 2021).

In 1907, the first plastic material was created, sparking the global plastics industry. However, the production of plastic didn't significantly increase until the 1950s (Hannah and Roser, 2018). The annual manufacturing of plastics expanded nearly 240 times over the following 70 years, going from 1.5 million metric tonnes to 490 million metric tonnes in 2022.

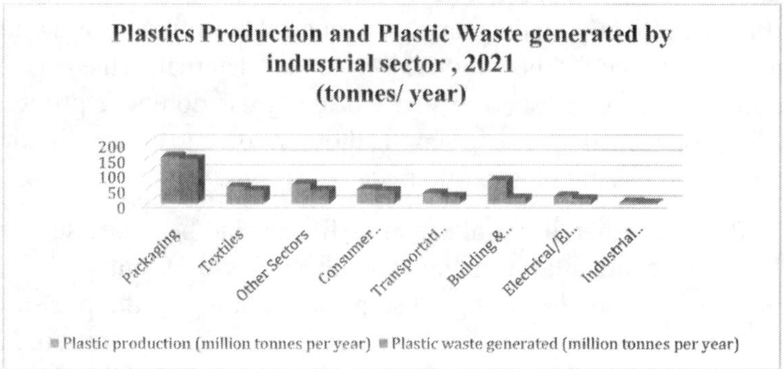

Fig. 1: Bar Graph showing the plastic production and plastic waste generated by industrial sector, 2021 measured in tonnes per year.

Source: European bioplastics, 2020.

Industries are also completely dependent on the plastics and packaging is the biggest industry which not only responsible for the maximum plastic production but maximum plastic waste is also generated by packaging industries. Other players are the textile industry and building and construction companies (European Plastics, 2022).

China is the world's largest producer and user of plastics, with a 60 million metric tonne annual consumption rate. The growth histories of many countries around the world show that as economies develop, so does the consumption of plastic goods. According to statistics, Americans use about 170 kilogrammes of plastic per person each year, as opposed to 200 kilogrammes in Belgium, 46 kilogrammes in China, and only 9.7 kilogrammes in India.

Ocean health- Sustainable development goal

The plastic issue poses a severe threat to our beautiful seas and marine life. Three-quarters of Earth's surface is made up of oceans, which also hold 97 percent of the planet's water and 99 percent of its liveable land. About 30 percent of the carbon dioxide that people produce is absorbed by the oceans, mitigating the effects of global warming (Watt et al. 2021). The ocean has also absorbed more than 90% of the additional heat in the climate system. Over three billion people rely on the marine and coastal biodiversity as a vital natural resource for their survival, which is what makes the oceans so important. A 5% share of the world's gross domestic product (GDP), or $3 trillion every year, is thought to be generated by the marine and coastal sectors and resources.

Plastic waste is harmful to marine life and has become a serious global environmental problem. The latest Covid-19 outbreak has intensified demand for single-use plastic, adding to the pressure on this already out-of-control problem. According to this study, there have been over eight million metric tonnes of plastic waste generated globally in connection with the epidemic, with over 25,000 metric tonnes ending up in the ocean (Stubbins et al. 2021). This is a long-term problem for the maritime ecosystem and largely accumulates on beaches and coastal sediments. According to the UN's Sustainable Development Goals, marine life and the health of the oceans are valued highly, and the government has started taking practical steps to conserve our oceans.

Plastic waste management routes

Plastic pollution is a major environmental concern, with approximately 390 million metric tonnes of plastic waste produced each year, 50 percent of which is single-use plastic and only nine percent of which is recyclable. The remainder of the plastics are unmanaged and simply land filled, where they will remain buried for the next thousand years due to their inability to degrade (Evode et al. 2021).

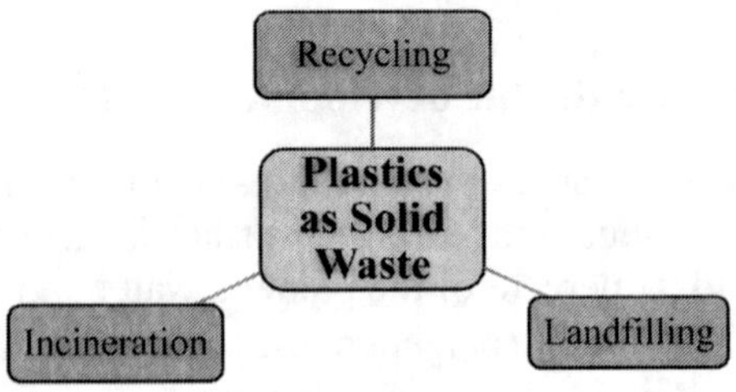

Fig. 2: Different routes of disposing plastic waste.

There are many novel trends in the recycling of plastics, but they are time-consuming processes that also contribute to the "greenhouse effect" in the environment. Mechanical, chemical, and biological recycling techniques are used these days (Idumah and Nwuzor, 2019). Microorganisms are also natural decomposers, which can help nature in the degradation of plastics. But these processes demand facilities and special composting units, which are not only time-consuming but also costly. Apart from that thermolysis technique is also there in which degradation process takes place at higher temperatures (Wang et al. 2019).

The following are the different techniques used in the plastic waste management.

Fig. 3: Plastic recycling techniques.

Source: Pan et al. 2020.

According to research from the Organization for Economic Cooperation and Development, more than 300 million metric tonnes of plastic waste are released into the environment each year (OECD). These plastic wastes have the potential to break down at precise temperatures and release toxins that harm the human liver, kidneys, and central nervous system (Wang et al. 2019; Cox et al. 2019). In the meantime, because most waste plastics are difficult to degrade organically, a sizable amount of plastic trash dumped into rivers and oceans disrupts the environment and causes the extinction of fish and other marine life. It is also well acknowledged that the purpose of waste plastic disposal is low-cost eradication as opposed to advantageous value-added reuse (Monteiro et al. 2018; Brebu, 2020).

Bioplastics as an alternative solution

Particularly since the 1950s, the plastics industry has grown at an incredible rate on a global scale. A push for more sustainable options has resulted from worries about the usage of petroleum and natural gas as plastics feedstocks and the environmental problems brought on by conventional plastic waste (Sakti et al. 2021; Dohler et al. 2022). Bioplastics are defined as plastics that are either biobased, biodegradable, or both. They are made from biomass resources that are renewable, like sawdust, vegetable oils, straw, and maize starch (Rahman et al. 2021). Only one percent of the plastic now produced worldwide is created from bioplastics. This is anticipated to increase as bioplastics become more complex and diverse. Global production of bioplastics is anticipated to reach 2.87 million metric tonnes by 2025, up from an estimated 2.11 million metric tonnes in 2020 (Nandakumar et al. 2021).

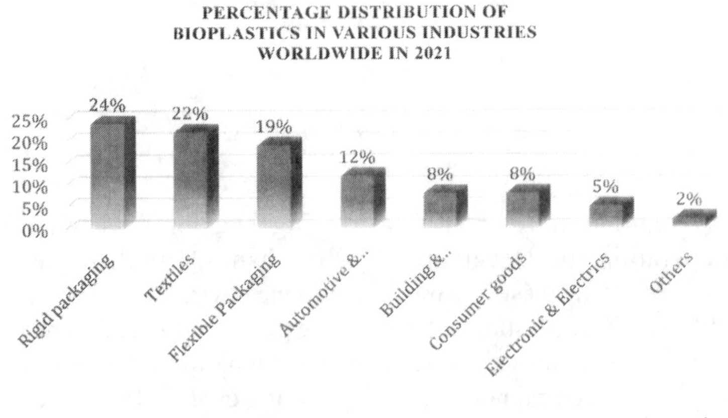

Fig. 4: Bar graph showing Percentage distribution of Bioplastics in various Industries worldwide in 2021.

Source: European bioplastics, 2021.

Nowadays, many bioplastics are available on the market. They are mostly plant-based, and because they are obtained from natural resources, bioplastics can be biodegradable and reduce carbon footprints in the atmosphere. 24 percent of the rigid packaging industry uses biodegradable polymers, and many industrial sectors started using these materials just to minimize the use of conventional plastics (European plastics, 2021; Cucine et al. 2021).

The data shows the percentage distribution of bioplastics in various industries for the year 2021. Still, there is scope for improvement in the production and quality of the bioplastics to maximise their usage in various other sectors, including commercially.

Bioplastics in promoting circular economy

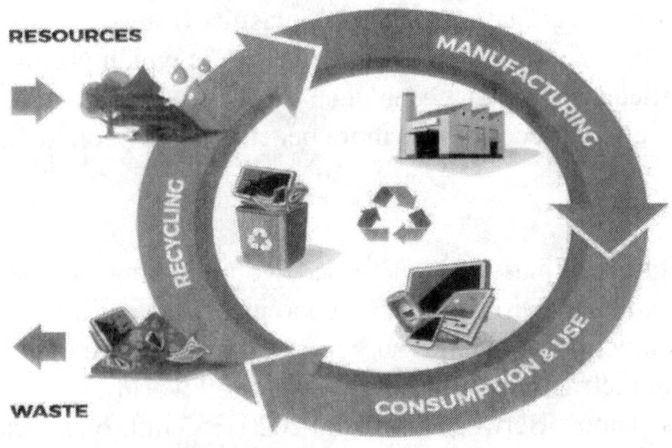

Fig. 5: Bioplastics Circular economy.

Source: Green dot plastics.

Utilising renewable resources by itself does not ensure sustainability. Sustainability is more influenced by a material's production method, intended use, and recycling capacity than by any of its component elements. Bioplastics, however, have the potential to drive a number of plastic-intensive firms in the direction of a circular economy as a result of technological improvements

(Rosenboom et al. 2022; Karan et al. 2019). Bio-based alternatives can almost always be used in place of fossil fuel,but they often come in small, expensive quantities and don't necessarily have major environmental benefits (Bartolo et al. 2021).

The three parts of the circular economy are system effectiveness, resource yield optimization, and preservation and enhancement of natural capital. The circular economy's success depends on these three elements (Rosenboom et al. 2022). Bioplastics are the primary players in the plastics industry's shift from a wasteful linear economy to a circular one. The New Plastics Economy will benefit from their expanded use in both functional and environmental ways, as well as in terms of the economy and the environment (Visco et al. 2022). Still, we are facing lot of problems in the production and degradation of these bioplastics as they demand special industrial composting techniques and for recycling process, there is a need for proper segregation of the bioplastics from the conventional ones. But as compare to linear economy in which plastic waste is land filled and unmanaged as such for years causing environmental issues, circular economy is more beneficial and boon to the nature.

Conclusion

The use of plastic has increased twofold during the past 30 years as a result of the growth of new economies. Between 2000 and 2020, the production of plastics surged by a factor of two, reaching 460 million metric tonnes, or 3.4 percent of all worldwide greenhouse gas emissions. Between 2000 and 2020, 353 million metric tonnes of plastic waste were generated globally. Nearly two-thirds of plastic garbage is made of polymers with a life span of less than five years; 40 percent of this waste is packaging, 12 percent is consumer goods, and 11 percent is clothing and textiles. Only nine percent of the plastic waste that is discarded is recyclable, despite the fact that 15 percent of it is collected for recycling; the remaining 40 percent is disposed of as residue. A further 19 percent is burned in landfills, 50 percent of it ends up there, and 22 percent of it manages to elude waste collection systems and ends up in uncontrolled dumps. The market for bioplastics is being driven by factors like rising consumer eco-awareness, mandates and regulations, corporate sustainability initiatives, technology stabilization, and cost-cutting. It is urgent to replace these conventional with biobased biodegradable polymers.

References

Brebu, M. (2020). Environmental degradation of plastic composites with natural fillers—a review. *Polymers*, *12*(1), 166.

Cox, K. D., Covernton, G. A., Davies, H. L., Dower, J. F., Juanes, F., & Dudas, S. E. (2019). Human consumption of microplastics. *Environmental science & technology*, *53*(12), 7068-7074.

Cucina, M., de Nisi, P., Tambone, F., & Adani, F. (2021). The role of waste management in reducing bioplastics' leakage into the environment: a review. *Bioresource Technology*, *337*, 125459.

Di Bartolo, A., Infurna, G., &Dintcheva, N. T. (2021). A review of bioplastics and their adoption in the circular economy. *Polymers*, *13*(8), 1229.

Dohler, N., Wellenreuther, C., & Wolf, A. (2022). Market dynamics of biodegradable bio-based plastics: Projections and linkages to European policies. *EFB Bioeconomy Journal*, *2*, 100028.

European plastics-(2022); https://plasticseurope.org/knowledge-hub/plastics-the-facts-2022/

Evode, N., Qamar, S. A., Bilal, M., Barceló, D., & Iqbal, H. M. (2021). Plastic waste and its management strategies for environmental sustainability. *Case Studies in Chemical and Environmental Engineering*, *4*, 100142.

Gall, S. C., & Thompson, R. C. (2015). The impact of debris on marine life. *Marine pollution bulletin*, *92*(1-2), 170-179.

Green dot plastics blog. https://www.greendotbioplastics.com/bioplastics-role-new-plastic-economy/

Hannah Ritchie and Max Roser (2018) - "Plastic Pollution". Published online at OurWorldInData.org. Retrieved from: 'https://ourworldindata.org/plastic-pollution'.

Idumah, C. I., &Nwuzor, I. C. (2019). Novel trends in plastic waste management. *SN Applied Sciences*, *1*(11), 1-14.

Karan, H., Funk, C., Grabert, M., Oey, M., & Hankamer, B. (2019).

Green bioplastics as part of a circular bioeconomy. *Trends in plant science, 24*(3), 237-249.

Lebreton, L.C.M.; van der Zwet, J.; Damsteeg, J.-W.; Slat, B.; Andrady, A.; Reisser, J. (2017). River plastic emissions to the world's oceans. *Nat. Commun., 8,* 15611.

Monteiro, R. C., do Sul, J. A. I., & Costa, M. F. (2018). Plastic pollution in islands of the Atlantic Ocean. *Environmental Pollution, 238,* 103-110.

Moshood, T. D., Nawanir, G., Mahmud, F., Mohamad, F., Ahmad, M. H., & Abdul Ghani, A. (2022). Sustainability of biodegradable plastics: new problem or solution to solve the global plastic pollution? *Current Research in Green and Sustainable Chemistry,* 100273.

Nandakumar, A., Chuah, J. A., & Sudesh, K. (2021). Bioplastics: a boon or bane? *Renewable and Sustainable Energy Reviews, 147,* 111237.

Oktavilia, S., Hapsari, M., Setyadharma, A., &Wahyuningsum, I. F. S. (2020). Plastic industry and world environmental problems. In *E3S Web of Conferences* (Vol. 202, p. 05020). EDP Sciences.

Pan, D., Su, F., Liu, C., & Guo, Z. (2020). Research progress for plastic waste management and manufacture of value-added products. *Advanced Composites and Hybrid Materials, 3*(4), 443-461.

Plastics- the facts (2021), Plastics Europe- Enabling a sustainable future; https://plasticseurope.org/knowledge-hub/plastics-the-facts-2021/

Rahman, M. H., & Bhoi, P. R. (2021). An overview of non-biodegradable bioplastics. *Journal of cleaner production, 294,* 126218.

Rosenboom, J. G., Langer, R., & Traverso, G. (2022). Bioplastics for a circular economy. *Nature Reviews Materials, 7*(2), 117-137.

Sakti, A. D., Rinasti, A. N., Agustina, E., Diastomo, H., Muhammad, F., Anna, Z., &Wikantika, K. (2021). Multi-scenario model of plastic waste accumulation potential in Indonesia using integrated remote sensing, statistic and socio-demographic data. *ISPRS International Journal of Geo-Information, 10*(7), 481.

Stubbins, A., Law, K. L., Muñoz, S. E., Bianchi, T. S., & Zhu, L. (2021). Plastics in the Earth system. *Science, 373*(6550), 51-55.

Thompson, R. C., Olsen, Y., Mitchell, R. P., Davis, A., Rowland, S. J., John, A. W., ... & Russell, A. E. (2004). Lost at sea: where is all the plastic? *Science, 304*(5672), 838-838.

Visco, A., Scolaro, C., Facchin, M., Brahimi, S., Belhamdi, H., Gatto, V.,

& Beghetto, V. (2022). Agri-food wastes for bioplastics: European prospective on possible applications in their second life for a circular economy. *Polymers*, *14*(13), 2752.

Wang, W., Gao, H., Jin, S., Li, R., & Na, G. (2019). The ecotoxicological effects of microplastics on aquatic food web, from primary producer to human: A review. *Ecotoxicology and environmental safety*, *173*, 110-117.

Watt, E., Picard, M., Maldonado, B., Abdelwahab, M. A., Mielewski, D. F., Drzal, L. T., ... & Mohanty, A. K. (2021). Ocean plastics: environmental implications and potential routes for mitigation–a perspective. *RSC advances*, *11*(35), 21447-21462.

15

The Stimulus of the Soil Health Card on Fertilizer Usage and its Relevance to Sustainable Development Goals

R. Jayapriya and *B. Keerthika*

Introduction

Agriculture relies on soil resources for crop cultivation. Soil resource is an eminent natural resource among the other natural resources on the earth because it acts as the medium for our livelihoods like food and shelter. Sustainable Development Goals -15 is to protect, restore, and advance sustainable forest management, desertification mitigation, prevent and reverse land degradation, and stop biodiversity loss. There is a need to educate farmers in SHC based nutrient management so that the programme main motto is 'demand-driven' rather than 'supply-driven'. Success or failure of a civilization depends on how well people understand how to manage natural resources like water, air and soil. Ultimately, it paves way forward that how to prevent their loss or degradation (Montgomery 2012). Chemical pollutants are one of the most serious dangers to the above-mentioned natural resources, either directly or indirectly.

So many measures have been taken worldwide to mitigate the damages caused by the human-made degradation of natural resources. Soil plays a vital role in agriculture, because there is no alternate medium for it in cultivation aspects. Dhopte (2017) highlighted that there are seven types of soil: hills, black soil, red soil, alluvial soil, desert soil, laterite soil, and terrain soils. In India, red soil covers the most of the land mass (105.5 million ha), whereas black soil covers 73.5 million ha. Whereas alluvial soil covers 58.4 million hectares, laterite soils cover 11.7 million hectares, desert soils cover 30 million hectares, and hills and terrain soils cover 26.6 million hectares (Nguyen et al. 2022). Stocking and Murnaghan (2013) reported that in developing countries, soil degradation is a growing issue in cropping, and decreased soil fertility harms agricultural production. According to Lal (2015), A significant barrier to raising agricultural production is soil degradation, which is characterised as a loss in quality and a reduction in ecosystem services and commodities. Gorlach et al. (2004) characterised the processes of formation and regeneration of soil.

Healthy soil has all of the elements required for crop cultivation, while soil lacking in one or more nutrients decreases crop yield and quality. Soil fertility was defined by Lehmann et al. (2003) as "the capability of the soil to give essential nutrients in sufficient proportions and amounts for plant growth throughout the growing season". As a result, the proportion as well as quantity of major and minor nutrients influence the overall health of the soil. In agriculture, soil health is crucial because it helps to assure long-term yield by optimising fertilizer usage and avoiding waste. Most farmers increase the amount of synthetic fertilizers they use on a daily basis to boost productivity without understanding the soil fertility state of their crops (Lehman et al. 2015). Makadia (2015) found that in Surat district, farmers with Soil Health Cards had a higher average yield (101.10 tonnes/ha.) than those without Soil Health Cards (95.70 tonnes/ha.) whereas in the Navsari district, it was 92.18 tonnes and 85.92 tonnes per hectare. Raju et al. (2015) conclusively demonstrated that the adoption of soil health cards boosted land utilization, altered farming practises in favour of crops that are additional profitable, and enhanced agricultural yields of the crop. The maximum and minimum salinity were significantly reduced. The total output of all crops planted grew by 186 percent on average. A rise in net income was mostly owing to an increase in output of the crop as a result of soil improvements. Off-season crops also resulted in a large increase in net revenue. The Benefit: Cost ratio was more than one. The value of land has also soared. Bordoloi and Das (2017) reported that cost of cultivation was reduced by Rs.1000 to Rs 4000 per acre. In soybean, overall cultivation costs were reduced by 1.7 percent from Rs.10900 to Rs 10714 per acre, but net revenue climbed by 67.7 percent from Rs 6696 to Rs 11228 per acre. After the farmers had their soil evaluated, the return on investment rose from 1.60 to 2.00 rupees. It is observed that the average net and gross income of paddy farmers in the Jorhat and Golaghat regions of Assam who used soil test-based fertilizers was 9282.50 Rs/ha and 41452.50 Rs/ha, respectively. The average gross and net revenue of paddy farmers in the research area who did not use soil test-based fertilizer recommendations was found to be 39755.00 Rs/ha and 7272.50 Rs/ha, respectively. He observed that the cost of cultivation for the control group was slightly higher than the cost of cultivation for the soil test recipient farmers. Shehrawat et al. (2018) determined that neither farmer had attended a soil testing facility or they are having a soil health card. Furthermore, it was shown that the majority of the farmers (61.33 percent) were aware of the SHC and believed that it assisted them in understanding the soil nutrient levels, doses of different fertilizer (45.56 percent), and finally crop selection in both the Kharif and Rabi seasons (52.22 percent). Kumar et al.(2019) found that the farmers have saved Rs 4,414 per hectare annually by following the fertilizer recommendations in the soil health card (in rice-wheat cropping pattern). Additionally, the condition of the soil improved over time, increasing agricultural production. In the cropping pattern (rice-wheat), the net returns and Benefit-Cost ratio rose by 19.8 and 16.1 percent, respectively. Production efficiency increased by 9.2 and 19.5 percent, respectively, while financial efficiency increased by the same amount. Senthamizhselvan et al.(2022) concluded that both SHC holders and non-beneficiaries had their overall costs calculated, with SHC holders' costs (Rs 17272.68/acre) being lower than non-beneficiaries costs (Rs 17710.32/acre) and they deduced that SHC Holders' cultivation costs (Rs 2717.66/acre) were lower than those of SHC Non-beneficiaries (Rs 28902.12/acre). The overall objective of this study is to evaluate the changes in fertilizer consumption and crop yield after adoption of Soil Health Card and to study about the constraints faced by the farmers.

Materials and Methods

Karaikal region of Puducherry U.T. was selected purposively to study the adoption of a Soil Health Card. In this study, a multistage random sampling method was employed. Twenty respondents with a Soil Health Card and twenty respondents without a Soil Health Card were chosen from each of the six firkas in the Karaikal area, making a total sample of 120 SHC beneficiaries and 120 non-beneficiaries. Table 1 shows how the sample farmers were distributed among the firkas.

Table 1: Details on Sampling of the Study

S. No.	Firka Name	Number of farmers	
		SHC Beneficiaries	**Non-Beneficiaries**
1.	T.R. Pattinam	20	20
2.	Neravy	20	20
3.	Thirunallar	20	20
4.	Nedungadu	20	20
5.	Karaikal	20	20
6.	Kottucherry	20	20
	Total	**120**	**120**

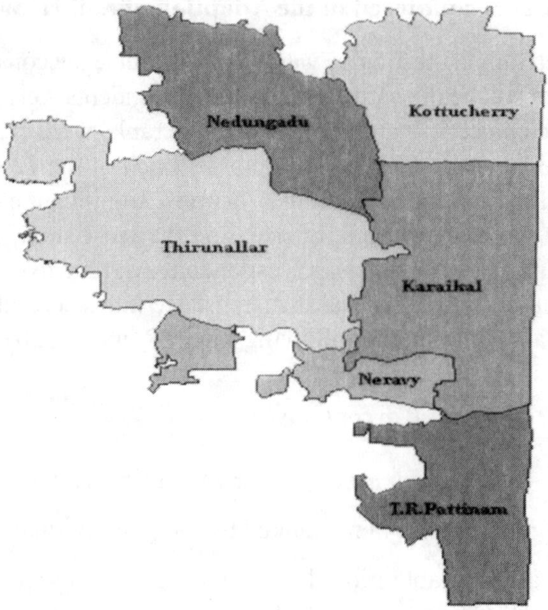

Fig. 1: Study area- Karaikal region of Puducherry U.T.

Benefit-cost ratio

The benefit-Cost Ratio is the ratio between the gross income and the cost of cultivation for crops.

$$\mathbf{BCR} = \frac{\boldsymbol{Gross}\ \textbf{Income}}{\boldsymbol{Total}\ \textbf{Cost}}$$

Where,

Gross Income=Income from the main product and by-product in one cropping season.

Total Cost = Expenditures incurred for inputs used as well as agronomic operations.

The BCR method was employed to assess the costs and advantages of crops grown in the study region for SHC beneficiaries and non-beneficiaries.

Constraints encountered in the Adoption of Soil Health Card

Garrett ranking technique was used to evaluate the constraints in adoption of soil health card scheme. The respondents were enquired to rank the constraints and these respective ranks were changed by using the Garrett table to convert points. Each of the 120 farmers was asked to rank the constraints. The most significant element in this study was given a rank of one, and the least significant was given a rank of n. The rank each respondent awarded to each factor under limits and strategies was then translated into a percent position using the algorithm in the following stage (Ramarao, 2011):

$$Per\ cent\ position = \frac{100\,(R_{ij}-0.5)}{N_j}$$

where, R_{ij} = Rank given for the item by the jth individual

N_j = Number of items ranked by the jth individual

By using the table provided by, the percent position of each rank was transferred to scores (Garrett and Woodworth, 1969). The scores of each respondent were then added together and divided by the total sample respondents whose scores were collected. Assigning ranks 1, 2, 3, etc. in decreasing order of mean scores resulted in the ranking of the mean scores for each factor.

Results and Discussions

Results from the Figure 2 clearly depicts that the fertilizer consumption was reduced after adoption of SHC. Nitrogenous fertilizer consumption was reduced by around 39 percent, consumption of phosphorous fertilizer was reduced by 12 percent and potassium was reduced by around nine percent after adoption of Soil Health Card. This resembles sustainable future environment for future generations and supports Sustainable Development Goal-15 by restoring land by reducing fertilizer consumption and increasing manure application. Similar results were obtained by Makadia (2015) concluded that the extent of N over-utilization in the South Gujarat region was found to be 2.94 percent for farmers with Soil Health Cards and 15.00 percent for farmers without Soil

Health Cards. For farmers with Soil Health Cards, the K fertilizer gap ranged from 9.09 percent in Navsari region to 10.98 percent in Surat district.

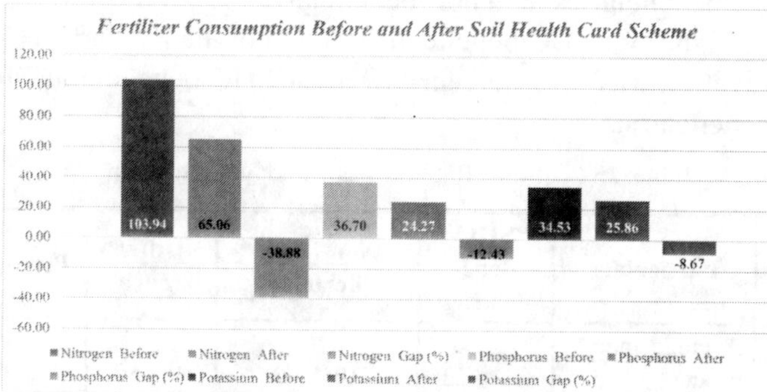

Fig. 2: Fertilizer consumption before and after SHC adoption.

Application of manures plays an important role in agricultural crop productivity. The consumption of FYM before and after the scheme's adoption in paddy cultivation, the gap between the actual and recommended usage of FYM for paddy was estimated and shown in Figure 3. Results from Figure 3 represents that manure consumption was increased to conserve soil resource and maintain sustainability as stated in SDG-15 i.e. land restoration and reverse land degradation.

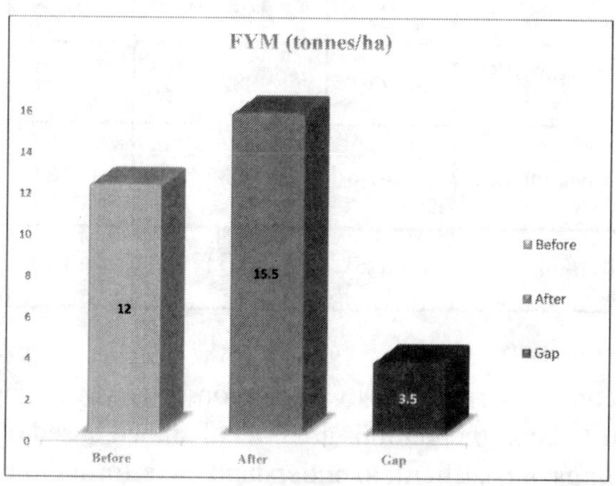

Fig. 3: Consumption of Manure before and after SHC adoption.

Cost and returns incurred and received during crop cultivation by the SHC beneficiaries and non-beneficiaries were depicted in the Table 2. Average main product yield was 9.33 percent higher for SHC beneficiaries than non-beneficiaries. B:C ratio was 13.20 percent higher for SHC beneficiaries than non-beneficiaries. Cost of cultivation was 4.75 percent lower for SHC beneficiaries than non- beneficiaries.

Table 2: Cost and Returns from Paddy Cultivation

Sl. No	Particulars	SHC Bene-ficiaries	Non-bene-ficiaries	Diffe-rence	Percent
1	Average main product yield (Kg/acre)	2116.23	1918.83	493.50	9.33
2	Average Sale Price (Rs /kg)	23.30	23.30	0.00	0.00
3	Returns from main product (Rs /acre)	49308.2	44708.8	4599.42	9.33
4	Returns from by product (Rs /acre)	2459.57	2357.65	101.92	4.14
5	Gross return (Rs./acre)	51767.8	47066.4	4701.34	9.08
6	Net return (Rs /acre)	14578.1	9082.72	5495.38	37.70
7	Cost of cultivation (Rs /acre)	36261.6	37983.7	-1722.1	-4.75
8	BC Ratio	1.43	1.24	0.19	13.20

Source: Primary survey (2021).

The constraints reported by the respondents while adopting soil health card, recommendations are not calculated based on farmers land holdings, it is difficult to understand the information provided on Soil Health Card, SHC reports are not delivered in right time to farmers, lack of NPK fertilizer and its various forms, inaccessibility of micro-nutrient fertilizer in the market, unaware of the grid system in collection of soil sample, soil test is not performed in farms with farmers' decision, unavailability of organic manure, insufficient knowledge regarding the significance of micronutrients and lack of government subsidy. Garrett ranking results were given below in the Table 3.

Table 3: Constraints encountered by SHC farmers

Constraints encountered by SHC farmers	Percent	Rank
Recommendations are not calculated based on farmers land holdings	73.97	1
It is difficult to understand the information provided on Soil Health Card	66.02	2
SHC reports are not delivered in right time to farmers	60.44	3
Lack of NPK fertilizer and its various forms	53.25	4
Inaccessibility of micro-nutrient fertilizer in the market	50.39	5
Unaware of the Grid system in collection of soil sample	49.17	6
Soil test is not performed in farms with farmers' decision	41.36	8
Unavailability of Organic Manure	39.14	7
Insufficient knowledge regarding the significance of micronutrients	35.11	9
Lack of government subsidy	17.43	10

Source: Primary survey (2021).

From the Table 3, it shows that majority of the farmers stated that recommendations are not calculated based on farmers land holdings was the major constraint faced by them while adopting SHC. A similar results were obtained by Kumar and Banga (2018).

Conclusions

From this study it is concluded that Soil resource should be conserved as requested in the Sustainable Development Goal-15 as its aim is to restore the land masses and reverse land degradation. Among the India's various primitive measures, Soil Health Card programme is resource conservation pathway. Fertilizer usage should be as demand driven and it paves way for sustainable environment. It is suggested that policy measures can be taken by the government to promote organic farming by giving finacial assistance to the farmers during the transition period from inorganic to organic crop cultivation.

References

Bordoloi, J., & Das, A. K. (2017). Impact of soil health card scheme on production, productivity and soil health in Assam. *Study No. 148. Study Sponsored by the Ministry of Agriculture and Farmers'* Welfare Government of India, New Delhi.

Dhopte, A. M. (2017). *Agrotechnology for Dryland Farming* 2nd: Scientific Publishers.

Garrett, H., & Woodworth, R. (1969). Statistics in Psychology and Education, 5th Indian ed. *Bombay: Vakils, Feffer and Simons Private Ltd.*

Görlach, B., Interwies, E., & Landgrebe-Trinkunaite, R. (2004). *Assessing the economic impacts of soil degradation: a literature survey.* Paper presented at the 2nd International Workshop on Integrated Soil and Water Protection: Risks from Diffuse Pollution (SOWA).

Kumar, A., Brar, N. S., Kumar, B., & Verma, H. (2019). Impact of fertilizer recommendations based on soil health card on fertilizer consumption, productivity and profitability of farmers. *Journal of Experimental Biology and Agricultural Sciences, 7*(3), 249-254.

Kumar, B., & Banga, G. (2018). An Empirical Investigation of Farmer's Perception and Attitude Towards Soil Health Testing in Punjab. *Journal of Agricultural Development and Policy, 28*(1), 81-91.

Lal, R. (2015). Restoring soil quality to mitigate soil degradation. *Sustainability, 7*(5), 5875-5895.

Lehman, R. M., Acosta-Martinez, V., Buyer, J. S., Cambardella, C. A., Collins, H. P., Ducey, T. F., . . . Kremer, R. J. (2015). Soil biology for resilient, healthy soil. *Journal of Soil and Water Conservation, 70*(1), 12A-18A.

Lehmann, J., Pereira da Silva, J., Steiner, C., Nehls, T., Zech, W., & Glaser, B. (2003). Nutrient availability and leaching in an archaeological Anthrosol and a Ferralsol of the Central Amazon basin: fertilizer, manure and charcoal amendments. *Plant and soil, 249*(2), 343-357.

Makadia, J. J. (2015). *Regional Imbalances and Impact of Soil Health Card on Fertilizer Consumption in Gujarat*: GRIN Verlag.

Montgomery, D. R. (2012). *Dirt: The erosion of civilizations*: Univ of California press.

Nguyen, T. T., Pham, T. D., Nguyen, C. T., Delfos, J., Archibald, R., Dang, K. B., . . . Ngo, H. H. (2022). A novel intelligence approach based active and ensemble learning for agricultural soil organic carbon prediction using multispectral and SAR data fusion. *Science of the Total Environment, 804*, 150187.

Raju, K., Gaur, A., & Wani, S. (2015). Poverty reduction approach in South Asia: Rejuvenating centuries old water bodies to improve rural livelihoods, A case of Karnataka state, India.

Ramarao, I. V. Y. (2011). An economic appraisal of manufacturing and marketing of jaggery in Andhra Pradesh state, India. *Sugar tech, 13*(3), 236-244.

Senthamizhselvan, D., Kumaravel, K., & Baskar, D. C. (2022). Impact of soil health card on fertilizer consumption and yield of paddy in Karaikal district for sustainable agriculture. *Seed, 1098*, 1069.1033.

Shehrawat, P., Aditya, A., & Sharma, N. (2018). Usage of soil health card in crop management practices for doubling the farmers' income. *Journal of Applied and Natural Science, 10*(4), 1314-1317.

Stocking, M. A., & Murnaghan, N. (2013). *A handbook for the field assessment of land degradation*: Routledge.

16

Survival of the Destitute in Kerala: A Grassroots Level Plan by Kerala Institute of Local Administration (KILA)

Nijil Jacobi

Introduction

Kerala is the region where many models were created before the world. The Kerala model of development refers to the practises followed in Kerala, which are distinguished by strong social indicators like high literacy, improved access to healthcare, high life expectancy, low infant mortality, and low birth rate—indicators that are frequently comparable to those seen in developed countries despite having a lower per capita income. These accomplishments, as well as the factors that contributed to them, have been identified as typical Kerala model outcomes. The state, which has launched programmes such as absolute literacy, family planning, and Kudumbashree, is now preparing for a historic achievement: identifying destitute and devising a plan for their survival. A population, free from poverty, is a very important factor in development standards. Kerala has also met human resource indicators such as high living and educational standards, as in developed countries. However, the state cannot be proud because extreme poverty exists in the midst of development and progress. Preparing survival plans and execute successfully is very urgent and significant in this contest. The present flagship programme for the destitute is a continuation of many government programmes aimed at the poor, such as Ashraya, Hunger Free Kerala, and Destitute Free Kerala. One of the government's top priorities is to eliminate extreme poverty within the next five years. The main goal of the programme is to identify destitute who should have been covered by the Ashraya scheme but were not, and to provide aid and programmes to help them to come out of the vicious circle of poverty. Thus, it is an important government project aimed at eradicating extreme poverty.

Discussion

We shall look into the planning strategy of Government of Kerala in this regard to eradicate the extreme poverty from the state and build a destitute free state. This research paper discusses many aspects of identification of destitute and planning proposals initiated by Kerala Institute of Local Administration (KILA) such as distinction between extreme poverty and poverty, prior interventions for the extreme poor by the state, participatory process to identify the extreme poor, advantages of participatory process, Local Self-Government Institutions (LSGIs) and poverty alleviation, objectives of the new scheme for the destitute, various stages of project implementation, and from extreme poverty determination process to micro-plan.

Extreme poverty and poverty

The term "poverty" is frequently used. Poverty is defined by the various types of disabilities that humans face like deprivation. Poverty was initially measured on the basis of food deprivation and caloric intake from food. That was a consideration when compiling the list of people living below the poverty line (BPL). However, rather than just food availability, the calculated indicators include clothing, housing, education, health, access to services, infrastructure, basic income, social poverty, and inaccessibility of welfare measures. Human Development Index (HDI) of United Nations Development Programme (UNDP) has become a primary index for accessing human resource development using the aforementioned life expectancy, education, and per capita income indicators. Kerala is a leading state in human development indices. But the studies show that the means the scars of poverty still remain in Kerala.

It is a trend to see poverty and extreme poverty are often lumped together in discussions. It acts as a barrier to resolution of extreme poverty by giving more emphasis on it. If poverty is defined as a general lack of capabilities, then the extreme poor are those who face the most severe conditions. The poor frequently face problems in obtaining basic necessities like food, clothing, and a safe place to live. But they always do not face difficulties in survival. A struggling family with little money to buy groceries on non-working days falls into the poor category mentioned above. They may also have difficulties in preparing the safe living space perfectly. But they may have basic accommodation. The extreme poor are a vulnerable group for whom meeting basic needs such as food, clothing, safe shelter, and a basic income is extremely difficult.

The following conditions can be seen as characteristics of the extreme poor,

- Those who are unable to overcome their condition by convincing others.
- Those who lack the physical and mental capacity to earn a living and survive.
- Daily those who do not have the ability to move their lives daily forward due to mental challenge, old age, chronic illness etc.

In general, the poor with employment capacity, basic income and basic living conditions are not the extreme poor. Destitute and destitute families should be considered extremely poor if they are unable to obtain even basic rights and are helpless, unable to go on with their lives without external assistance.

Families or people living on the streets, families with only categories such as surviving without other means of income or assets, the differently-abled, the elderly, physically and mentally challenged in crisis, families without a breadwinner (died or left) and no regular income source, families without opportunities, assets or work capacity, families with inpatients only and having no working family members and no other assets or income, are beyond absolute poverty and are extremely poor with no means of survival.

If we distinguish the poor and destitute, a family with relatively small house or limited space to each individual can be considered as poor. Whereas, a family wandering the streets with no place, no home or shelter is destitute. A family with basic living, availing food grains through Antyodaya Anna Yojana (AAY) scheme and has family members employed through employment guarantee scheme, is considered to be poor. At the same time, a family without ration card to access food grains and unable to take up employment is destitute. A family with inpatients, but family members taking employment is poor. Whereas, a family where one member is able to work but is unable to do so because of taking care of family member or members who are in sick beds. A poor family with elderly people can be considered to be a poor family. But a poor family with only elderly people and no one is able to go for work is said to be destitute.

The most severe form of poverty is extreme poverty. Those living in extreme poverty cannot advance in life or get out of poverty without the strong support of the government and general society on the outside. The ordinary poor, on the other hand, have income and work capacity and can progress, though somewhat gradually over time. In general, the government employs various methods to get the poor out of the poverty. The employment guarantee schemes to provide employment and income, ration system to provide food and housing schemes for shelter are all part of this. These aids generally enabling the poor to survive. But for various reasons, families are in dire straits without access to even these basic services. They may be at risk of extreme poverty.

Prior Interventions for the Extreme Poor

To find the very poor and needy, and help them to live many projects have been implemented in Kerala. The following are the important interventions for the extreme poor by the Government of Kerala.

Kudumbashree Mission

The Kudumbashree programme to end poverty and empower women is being carried out by the State Poverty Eradication Mission (SPEM) of the Government of Kerala. The name Kudumbashree refers to "prosperity of the family" in Malayalam. The name stands for "Kudumbashree Mission" or "SPEM" as well as the Kudumbashree Community Network. The Kudumbashree Mission, the Kudumbashree Community Network, or both may be referred to as "Kudumbashree". Kudumbashree is essentially Kerala's state-wide community network. In this three-tier organization, which also includes Area Development Societies (ADS) at the ward level and Community Development Societies (CDS) at the local government level, Neighbourhood Groups (NHGs) serve as the primary level units. It is one of the largest women's networks in the globe. Even though the community network's main objectives are the eradication of poverty and the empowerment of women, democratic leadership and support networks created from the "Kudumbashree family" are some of its most distinguishing features.

Ashraya

Ashraya is a comprehensive effort with the goal of locating and helping poor families. The Kudumbashree effort was continued in 2002 with the goal of identifying the families who had been overlooked by decentralised planning and poverty alleviation initiatives. According to the plan, each Grama Panchayat must develop unique micro projects for each recognised poor household. The Grama Panchayats' annual plans are then merged with these micro projects. The Grama Panchayat plan funding, Block and District Panchayat contributions, centrally sponsored initiatives, and State government funds are all combined to fund the programme.

101 Grama Panchayats in the State developed initiatives including the rehabilitation of 8233 impoverished families during the first year of implementation. The simultaneous field implementation of numerous micro plans was made possible by the institutional frameworks that the Grama Panchayats were able to secure.

By 2015, Ashraya had been put into practise in 32 urban local bodies and 890 Grama Panchayats. Projects totaling Rs 4107.6 million have been prepared for a total of 72,116 families. Thousands of initiatives are currently being carried out around the State.

Rehabilitation for the homeless involves a process-oriented approach under Ashraya. The Kudumbashree community organization has been instrumental in the scheme's ability to produce a number of success stories. The following unique qualities of the system are largely responsible for its successes:

- Destitute families are the project's beneficiaries, not individuals.
- The Grama Panchayat finds the poor family through Neighbourhood Groups; there are no applications to fill out or approval processes.
- If there are no surviving members of the poor family, the house built for them becomes the property of the Grama Panchayat, which may then give it to another family in need.
- Ashraya projects are reviewed through a special system at the District Planning Committee (DPC) level.
- Every Ashraya Project is co-financed by the State Poverty Eradication Mission using funds from the State Plan, up to a maximum of 25 percent of the project's cost, or Rs 10 lakh.

Destitute Free Kerala

When the Ashraya project, which had gone through multiple phases, was reorganised in 2017, a more comprehensive one-time endeavour called Destitute Free Kerala, or Agathirahitha Keralam, was developed. In contrast to earlier efforts to eradicate poverty, Destitute Free Kerala tries to study several facets of poverty rather than focusing on just one. Many of the basic, developmental, and psychological needs of a Destitute Free Kerala beneficiary family are taken care of under the auspices of Kudumbashree, but other needs are taken care of by government organizations like Local Self-Government Institutions (LSGIs), the Health Department, the Public Distribution System, etc.The funding for the various project components comes from projects and plans that are already in operation. Kudumbashree, LSGIs, and generous individuals and groups are funding the project. The Challenge Money is a fund that Kudumbashree allots for essentials. It accounts for 40 percent of the total project cost, with a maximum of 40 lakh for general projects and 50 lakh for ST projects. The lowest tiers of project monitoring are handled by the foot troops of the Neighborhood Groups of Kudumbashree.

Distitute Free Kerala, as contrast to Ashraya, uses better technology for project planning and recipient identification (using a mobile app), which makes the programme more efficient and timely. An appeal can be made, and appeals are reviewed by LSGIs. At the State and District levels, monitoring committees will be set up to keep an eye on the initiatives after they have been implemented. Families that don't meet the project's eligibility conditions but are jobless even 10 days a month will be given support under the category of Special Cases. Additionally, a clause that can be added at any time is included. Thus, it is predicted that Ashraya will more effectively handle the various facets of poverty now that it is renowned as Kerala's Destitute Free State.

Rehabilitation for the homeless involves a process-oriented approach under Ashraya. The government adopted "Agathirahitha Keralam", the Destitute-Free Kerala programme, a modified version of the Ashraya scheme after 2016. The government envisaged a comprehensive plan for the extremely poor through this programme, who for a variety of reasons are not covered by any of the aforementioned programmes. The Kudumbashree community organization has been instrumental in the scheme's ability to produce a number of success stories.

Participatory Process to Identify the Extreme Poor

It is said that extreme poverty is a condition where there is no ability to survive as indicated. They are not even able to speak up for themselves or advocate for themselves. There will be families or people those who suffer due to different types of social isolation. A family that was isolated by society due to mental challenges is an example for the same. Street vagrants don't have a home they can call their own. People who reside in slums or public places are not considered as part of a family or of the community. Empirically identifying the extremely poor via conventional methods may not be successful because they frequently go unnoticed and are unable to speak up for their own rights.

Use of a participatory process including a cross-section of the community and the poorest families at the local level with local support is therefore crucial. As a result, it enhances the significance of clusters and ward-level linkages within the ward where it is common to find extremely impoverished households. For this reason, it's critical to use the technique of locating the extremely poor through participatory discussions with representatives of diverse groups that are constantly present in society. In fact, through this participatory discovery process the entire community is involved in locating the extremely impoverished and providing for their survival.

Through a participatory discussion with each indicator of extreme poverty at and below the ward level, the difference between poverty and extreme poverty was made. Thus, the participatory identification of the extreme poor is possible with greater precision. This participatory process enables decision-making by recognising the extreme nature of extreme poverty beyond the general nature of poverty.

Advantages of Participatory Process

The following are the major advantages of the ward level or community based participatory process put forward by the Kerala government in finding the destitute and solving the problem of extreme poverty from the state.

1. This participatory process plays a key role to identifying the most deserving extreme poor.
2. It is possible to find those who are desperate and needy who usually go unnoticed.
3. It helps to recognise precisely the deserving people by distinguishing between poverty and extreme poverty.
4. As a participatory process, it becomes a general social approach to extreme poverty and its solutions by entrusting responsibilities to all the members of a community.

Local Self-Government Institutions and Poverty Alleviation

In Kerala, Local Self-Government Institutions (LSGIs) have been essential in efforts to reduce poverty. The project operations have greatly aided the poor, who have been receiving basic services from local government institutions like housing, drinking water, sanitation, and electrification for more than past 25 years. Along with this, special emphasis has been given to combat poverty in the sub-projects for Scheduled Tribes and Scheduled Castes. But extreme poverty was addressed with the ASRAYA scheme of 2002-2003. There were certain interventions to formulate programs at the local government level. Then an anti-poverty sub plan was introduced to coordinate all kinds of poverty eradication efforts of the government.

The Kudumbashree project, which aims to reduce poverty, was launched in Kerala in 1998. It is a pivotal moment in Kerala's history. Following the projects, Kudumbashree and local self-government organizations participated in every initiative of poverty alleviation. Kudumbashree is also in charge of the Ashraya project. Additionally, a local self-government body's sub-scheme for reducing poverty was put into action. The extreme poor are the focus of the "Agati Rahitha Keralam" (Destitute-Free Kerala) scheme, which was introduced after 2016. The destitute-free Kerala initiative placed a strong emphasis on the requirements for fundamental living conditions, survival, and development. The new project is a continuation of the various projects implemented to eradicate extreme poverty. A new approach that incorporates the extremely poor—who were either left out of poverty alleviation efforts or weren't given much attention in earlier programmes—is supported for a number of reasons.

Objectives of the new scheme for the destitute

The new scheme for the destitute aims at the following objectives.

1. Eliminate extreme poverty in the state within five years.
2. Identify those experiencing extreme poverty for various reasons through the social participation process in every Local Self-Government area of the State.
3. Carry out the collection of comprehensive data on the status of each extreme-poor family.
4. Prepare family-based micro plans to enhance the livelihood of survival and the standard of living for the poorest families.
5. Prepare and implement local body level plan by coordinating micro plans.

Various stages of project implementation

The current project has to be implemented gradually in a continuous process. There has to be a well-organised strategy to achieve the stated goal of destitute free Kerala. The following are the proposed stages of the plan.

1. The idea of the micro plan has to be given to the public representatives, officers and volunteers through training sessions regarding the process and methodology of identifying the extreme poor.
2. People's Committees have to be formed at Local Self-Government level and Ward level.
3. Formation of state level coordination committee and district level committee for project coordination.
4. Preparation of preliminary list of extreme poor.
5. Conducting primary data collection based on fixed criteria in listed families.
6. Super checking of data collection process to distinguish poor and destitute.
7. Compilation of collected data and preparation of ward-wise list of extreme poor.
8. Preparation of the list of eligible persons and approval of local body.
9. Presentation and approval of ward level lists in Ward/Grama Sabhas.
10. Preparation of final list of eligible persons at local body level.
11. Planning and preparation of Local Self-Government level scheme and separate micro plan for each family.
12. Acquiring and Mobilising necessary resources for the implementation of Local Self-Government level plan and micro-plans.

From extreme poverty determination process to micro-plan

The government has taken up the big goal of eradicating extreme poverty in five years. The remarkable decision was taken in the first cabinet meeting of the new Kerala government that came to power in 2021. A wide-ranging action to eradicate extreme poverty at the local level was organised under the leadership of Local Self-Government Department and Kerala Institute of Local

Administration (KILA) in 2021. State-level extreme poverty determination cell headed by Development Commissioner, Rural Development Department worked for this purpose. Extensive training programmes from state level to ward level were organised for this initiative. More than 4,00,000 people were trained and over 60,0000 focus group discussions in the state were organised as part of this process. More than 13 lakh people participated in this project. The process of identifying the extremely poor was done through the focus group discussions held at ward levels. The lists of destitudes prepared in this way were approved by the Gram Sabha. Finally, it is discovered that 64006 families are destitudes as a result of extreme poverty. The process of determining extreme poverty is based on the major indicators such as food, basic income, health and housing. Along with this, special treatment was given in the selected indicators to include the Scheduled Castes, Scheduled Tribe families, coastal dwellers, etc. Special consideration is also given to poor families belonging to HIV victims, families protecting orphaned children, LGBTQIA+ categories and urban poor.

Conclusion

The United Nations' Sustainable Development Goals (SDGs) state that their top priority is to end all forms of poverty by the year 2030. It also aims to increase access to key services, give the poor and vulnerable social protection, and help individuals who have been impacted by natural disasters, economic shocks, and extreme weather conditions. The government of Kerala has proposed a flagship initiative complete eradication of extreme poverty from the state, which needs to be accomplished within five years, in this strict Covid-19 scenario, when the globe attempts to reduce the amount of poverty into the half percent of the current level. The government of Kerala has started the initial stages right away with a grassroots participatory strategy engaging local self-governing institutions and the general public.

The most critical step ahead is to prepare a micro plan for the families identified through the extreme poverty identification process. Micro plan has to be prepared by identifying the situation of each family and their immediate survival needs for both short-term and long-term activities. And from where each of these will get resources should be included in the micro plan. The committees formed at the local body level are responsible for preparation of micro plans. Local bodies also aim to coordinate family-based micro-plans and prepare local body level sub-plans. In this regard, with the intention of alleviating extreme poverty, Kerala Institute of Local Administration (KILA) of the Government of Kerala began the training programme by developing comprehensive strategies to assist the poorest households by raising their standard of living. KILA began training Local Self Government officials to create micro plans in order to improve the lives of the families that have been designated as the poorest in the state.

References

Complete eradication of extreme poverty in Kerala in next 4 yrs: Minister, 2022, July 26, *Business Standard*.https://www.business-standard.com/article/current-affairs/complete-eradication-of-extreme-poverty-in-kerala-in-next-4-yrs-minister-122072600056_1.html

Kerala Institute of Local Administration Training to Eradicate Extreme Poverty, 2022, August 4, *The Times of India*.http://timesofindia.indiatimes.com/articleshow/93335530.cms?utm_source=contentofinterest&utm_medium=text&utm_campaign=cppst

KILA, Kerala Institute of local Administration (n.d.). https://www.kila.ac.in/

Kudumbashree, State Poverty Eradication Mission, Govt. of Kerala (n.d.).https://www.kudumbashree.org/

Preparation of the Micro Plan for Destitute. (July 2020). One-Day Training Module, Kerala Institute of Local Administration (KILA).

The Kudumbashree Story (n.d.). https://thekudumbashreestory.info/index.php

The Process of Finding the Destitute. (October 2021). Hand Book, Kerala Institute of Local Administration (KILA).

Training for local Body Heads in Extreme Poverty Eradication, 2021, October 28, *The Hindu*.https://www.thehindu.com/news/national/kerala/training-for-local-body-heads-in-extreme-poverty-eradication-held-in-hollam-district/article37210873.ece

17

Role of Caste in Access and Choice of Agricultural Credit

Suresh Govindapuram, Thottadi Bhavani Prasad and *P.V. Bhargav*

Introduction

Agriculture credit is a direct and indirect intervention in improving agriculture production and productivity. Direct in terms of short, medium and long-term credit and indirect in terms of getting subsidies for seeds, fertilizers and agricultural mechanization techniques (Chaudhuri & Cherical, 2012). Credit from formal financial institutions serves the consumption smoothing effect and liquidity effect, i.e., working capital and investment credit (Binswanger and Khandkher, 1992; Narayan, 2015). Institutional credit positively impacted agriculture productivity (Binswanger & Khandkher, 1992; Chavan & Ramakumar, 2007; Narayan, 2016). Binswanger and Khandker (1992) argue that institutional growth and agricultural credit lending growth helped increase crop output. More recent work indicates a 0.22 elasticity between agricultural credit and real agricultural GDP (Subbarao, 2012). Some farmers, especially small and marginal farmers, lack access to agriculture credit from formal or institutional sources of finance (Chaudhuri and Cherical, 2012).

Access to different resources and opportunities is vital for improving the ladder in the social set-up, particularly for the disadvantaged. Creating Equality in Political, Social and Economic opportunities for different sections of the people is very important for better social outcomes. But the underprivileged groups face several hindrances in accessing resources and getting opportunities. The hindrances don't necessarily come down over a period; it requires intervention from the state to make it more inclusive. Credit is one financial resource access; it improves economic mobility through social mobility. The lack of these resources for disadvantaged sections perpetuates the existing inequalities (Kumar & Venkatachalam, 2019).

Social group identity is a basicorm of identity that influences economic outcomes. A caste is an important identity in India, often affecting every state of life (Deshpande, 2000). Indian society is mainly stratified on caste, which historically influenced economic outcomes, has a substantial impact on employment and is associated with occupation (Thorat & Attewell, 2007) and consumption (Deshpande, 2000). Agricultural production is a continuous process involving different stages and requires finances to complete the process. Agricultural credit is an important instrument and an important option to finance agricultural production. It acts as a catalyst in the production process, between the initial phases and till harvesting. Access to credit determines the farmers' production level and, in turn, the standard of living. So, the pertinent question is whether caste plays any role in accessing agricultural credit, particularly institutional credit. Though studies are based on caste influence in accessing credit (Kumar, 2013; Burges & Pande, 2005), the most critical question is whether lenders discriminate while providing credit (Kumar, 2013). So, this paper examines the role of caste in accessing credit. Are there any differences in credit access among the different social groups? If there are, what could be the causes behind the differences? The differences in accessing credit levels might indicate discrimination. Thorat & Attewell (2007) study found the Existence of discrimination in the labour market. Likewise (Altonji & Blanka, 1999), a study finds discrimination based on gender and race.The same findings are identified by Asiedu et al. (2012) study. He found that the loan denial rate is higher for minority-owned, black-owned, and minority-owned industries that paid more interest rates than the majority groups. It is imperative to observe whether there is discrimination in credit lending, particularly formal lending. It is also necessary for India's financial inclusion and agriculture where 69.2 percent of the total small and marginal farmers in India comes under the disadvantaged groups (Census, 2011).

Background

Indian society is a hierarchal caste system. It emerged from the Hindu religion and identified people based on their occupation. Thus, occupations are hereditary, and their marriages are endogamous. Even though the occupation is not ascribed in the present-day context, it still influences obtaining physical, financial, and social capital (Vijayabaskar & Kalaiyarasan, 2014). The entire population in India is categorised into four significant categories (OC, OBC, SC, ST). OBC, SC and ST are treated as disadvantaged groups, though they vary in degree. SCs are considered untouchables, the most underprivileged historically, OBC are backward castes, relatively better off than SC, ST are tribals, historically not part of the Varna system (Ghurye, 1969; Das et al. 2003).

Indian society is a hierarchal caste system, and caste identity is crucial in every sphere of life. So, it is intrinsically related to economic outcomes.

After the independence, GOI came up with different affirmative actions to uplift historically disadvantaged groups. In 1950, the Government of India started the Planning Commission, the first five-year plan focused mainly on improving the agricultural scenario for self-sufficiency in agricultural production. On the credit front, the Reserve Bank of India, 1954 introduced an India credit committee to identify the factors which could determine financial accessibility. Committee found that socioeconomic status strongly influences financial accessibility (RBI, 1954). Later in 1969 few banks nationalised to expand one branch in every three villages to enhance rural development (RBI IWG, 2019) given providing enough credit to rural areas, GOI came up with the 1980 banks' nationalization, RRB establishment, and NABARD establishment. During the 1991 reforms, the central government formed the Narasimhan committee to increase banks' operational efficiency, thereby, credit. NABARDs 1999 Kisan Credit Cards for easy credit, doubling agricultural credit (2004), Interest subvention schemes (2006) for providing cheaper loans, and priority sector lending revised scheme (2012) played a massive role in providing rural credit and in its transformation. Recently in 2014, GOI also came up with Jan Dhan Yojana as part of its financial inclusion policy. Though governments develop different policies and plans, the interventions that enhance credit access are largely debated.

Burgess and Pandey (2005) say social banking, i.e., the expansion of bank branches in rural areas during the nationalization of banks, helped reduce poverty by providing easily accessible credit to rural areas. Along with the social banking scheme, priority sector lending, lead bank scheme, service area approach to increase, the differential rate of interest scheme to facilitate credit at concessional interest rates for the poor, evolution of SHGs to bank linkage programme for weaker sections, RIDF to increase the flow of credit to rural areas. These schemes positively impacted rural credit flow (Kumar et al. 2017; Mohan, 2006). Implementation of a doubling agricultural credit programme helped increase the supply of credit, and since then, agrarian credit flow has constantly surpassed the aimed target. Though there is a rise in the flow of formal credit to agriculture, its share of proportion to total credit was declining. Mehrotra (2011) argues that annual agriculture credit growth rates have increased in recent years, but the rise in credit is not inclusive. These negatively impact the small and marginal farmers' credit accessibility (Mohan, 2006; Mehrotra, 2011; Kumar, 2013; Fatchamps, 2013). Along with this Existence of regional disparities among various regions in the flow of credit (RBI IWG, 2019) and the existence of inequalities in credit access among different social groups (Kumar, 2013; Kumar and Venkatachalam, 2016; Rajeev et al. 2011). Unless the government solves the inequalities in credit accessibility, achieving double income farmers' income by 2022, will remain an illusion.

Disadvantaged groups are the most financially excluded. Enhancing their conditions is one of the policy concerns in India. But the pertinent question is whether the institutional lenders discriminate against farmers based on caste is the key. There are two ways to describe discrimination in the rational choice model, one is the t-based discrimination i.e., lenders' decisions to provide loans are based on subjective factors (Becker, 1971). And the other Arrows (1973) statistical discrimination is based on the lenders unobserved objective characteristics like credit worthiness, capacity to pay etc. Hence, the real question arises whether the disadvantaged face t- based or statistical discrimination. The disadvantaged groups in India, who are less endowed with financial and physical assets, their less collateral capacity acts as a constraint to getting loans. The real problem of discrimination arises when they have been given less amount even though they have the equivalent amount of assets to collateralise.

This paper considered three questions to know the differences. Farm households obtain credit from commercial banks, similarly for microfinance and cooperatives and finally for money lenders. As I am only to find the differences in accessing the credit, other variables are taken into consideration like cultivated area, aggregate assets, education of an adult, caste, and per capita income.

Data: This paper used data from IHDS (the Indian human development survey, 2012). This data consists of 42,152 sample households. This survey was conducted by researchers of the University of Maryland and the NCEAR (the National Council of Applied Economic Research), New Delhi. This data consists of social characteristics, consumption and living standards, energy use, agriculture information, income and employment, sources of debt, health, education, and village infrastructure. Data also gives more information on the loan application and approval rates separately and data on choice of loan sources credit such as formal sources (banks/cooperatives /govt.), semi-formal sources (microfinance/

SHG), money lenders, employers, friends and relatives, other sources. The nature of the loan process, and the sources, which are critical to examine the caste differences in loan patterns, do certainly mirror discrimination on the part of the lenders. This complete loan information gives an edge over the NSSO data to analyse the discrimination. IHDS also contains rich information about socio-economic and demographic features, fitted well for determining credit access.

This study covers the whole of India. Out of the 42,124 households, 26.45 percent OC, 43.82 percent OBCs, 19 & 9 percent of SC and STs, respectively. 53.72 percent of samples have taken loans from banks, and the rest did not. 21 percent of people took from commercial banks, 9.6 percent dependent on cooperatives and microfinance and 15.10 percent from money lenders. Given Agricultural land holding in the data is divided into three categories 1. Small and marginal (less than are equal to 5 Acres) 2. Medium (5 to 25 Acres) 3. Large farmers (more than 25 Acres) for this study. Household total assets, per-capita income, and adult education are also considered.

The present study focuses on access to formal and informal credit by the small marginal, medium and large farmers' social groups and among the various social groups. Examine caste differences and differences in the scale of accessing or obtaining credit in both formal and informal sectors. If differences exist in access to credit, it is an important question to be answered in that the former is more important than the latter.

Methodology and Results

Many previous studies used binary response models like logit and probit models for studying agricultural credit access. This paper modelled probit to find farm household credit choices, particularly for commercial banks, cooperatives-microfinance and money lenders. The choice of credit was seen for various social groups. Modelling probit for the study is an arbitrary choice, as logit and probit give almost identical results, i.e., if we scale the probit coefficients with 1.6, we can obtain logit estimates.

$$y_i^* = x_\beta' + e_i$$

$$And \quad y = 1 \ if \quad y_i^* > 0$$

$$e_i \sim N(0, 1)$$

In the above univariate model, Yi is the dependent variable, which is binary, i.e., the latent variable is observed or not; if the outcome is yes, we take the value as one; if no, we take it as 0. To know whether there are differences in accessing credit between small and marginal farmers, and medium and large farmers. And also the study has tried to understand the same for various caste groups, among the small, marginal, medium and small-scale farmers. As the dependent variable is binary, the study has used probit estimates. The probit function consists of loans as the dependent variable, where total assets, agricultural land, education, and caste (dummy) are taken as a dependent variable.

The above estimates are probit coefficients obtained through the Probit model. Only caste coefficients are mentioned above. Explains that unit change in the independent variable leads to changes in the z score of the dependent variable. A higher increase in z scores gives higher chances of obtaining credit.

Table 1 explains the loans of small-marginal, medium and large-scale farmers able to access credit from commercial banks. The table indicates how access to credit varies among the various caste groups. We investigate whether caste plays a role in accessing credit from commercial banks. In the case of small and marginal farmers, it indicates no differences among castes in access to credit, as values are insignificant. But when it comes to medium and large farmers, SC and STs are less likely to get loans than OBCs. The coefficients, i.e. when OBC medium and large farmers obtain a loan, the Z score increases by 0.22 and 0.34, compared to 0.15 and 0.27 of STs and -0.20 and -0.16. We can observe that there are inter-caste differences in obtaining the loan, and it is evident that SC and STs have relatively less getting to credit. Assets are essential for small and marginal farmers compared to medium and large. Assets play no role in obtaining credit because they already have large-scale lands. Per capita income is more critical for small and marginal farmers than for medium and large farmers with a higher probability of credit access.

Table 1: Credit obtained by households from banks

	(1)	**(2)**	**(3)**
Variables	**Small & Marginal**	**Medium**	**Large**
OBC	0.221***	0.211***	0.345***
	(0.0332)	(0.0434)	(0.0195)
SC	0.0491	-0.154**	0.263***
	(0.0417)	(0.0615)	(0.0227)
ST	0.000612	-0.204**	-0.161***
	(0.0493)	(0.0798)	(0.0344)
ASSETS	0.0312***	0.0237***	-0.00958***
	(0.00283)	(0.00394)	(0.00162)
INCOMEPC	2.18653*	0.51719	1.757.6*
	(0.59769)	(0.50920)	(0.13672)
HHEDUC	0.0183***	0.0171***	-0.00171
	(0.00325)	(0.00464)	(0.00201)
Constant	-1.313***	-0.892***	-0.0192
	(0.0466)	(0.0663)	(0.0281)
Observations	10,823	4,798	26,398

Standard errors in parentheses
*** indicates significance at 1 percent ($p<0.01$), ** indicates significance at 5% ($p<0.05$),
* Significance at 10% ($p<0.1$)

Inter-caste differences are observed for cooperatives and microfinance, regardless of the farmer scale. Differences in the likelihood chances in obtaining credit are visible. These marginal differences are significant in the farmer's scale and across caste groups. When an OBC is small, medium and large farmers, the

Table 2: Credit obtained by households from Cooperatives and microfinance

	(1)	(2)	(3)
Variables	**Small & Marginal**	**Medium**	**Large**
OBC	0.281***	0.305***	0.331***
	(0.0481)	(0.0688)	(0.0275)
SC	0.317***	0.326***	0.296***
	(0.0570)	(0.0900)	(0.0315)
ST	0.164**	0.204*	0.132***
	(0.0684)	(0.118)	(0.0480)
ASSETS	0.0174***	0.00644	0.00178
	(0.00402)	(0.00612)	(0.00222)
INCOMEPC	2.0812***	1.52872*	1.86268***
	(0.91878)	(0.91009)	(0.58206)
HHEDUC	0.00346	0.00873	0.00431
	(0.00440)	(0.00699)	(0.00266)
Constant	-1.801***	-1.838***	-1.435***
	(0.0657)	(0.103)	(0.0392)
Observations	10,823	4,798	26,398

Standard errors in parentheses.
*** indicates significance at 1% (p<0.01), ** indicates significance at 5% (p<0.05)., * Significance at 10% (p<0.1).

increase in Z score is 0.28, 0.30 and 0.33. In the case of small marginal, and medium farmers, in contrast to commercial banks, SCs have more likely access to credit due to a lesser increase in z scores while obtaining a loan from cooperatives and micro finances. I.e. for SC, the change in z score is higher, i.e. 0.31, 0.32 for small and medium households. STs have less difference in Z score compared to the other two sections. Small and marginal farmers face less likely chances of obtaining credit than medium and large farmers. Assets matter for small and marginal farmers compared to medium and large farmers in obtaining credit from cooperatives. Farm size does matter in obtaining credit; the higher the farm size better the probability of getting credit. It is similar across the caste groups.

Table 3: Credit obtained by households from Money lenders

	(1)	(2)	(3)
Variables	**Small & Marginal**	**Medium**	**Large**
OBC	0.391***	0.331***	0.464***
	(0.0403)	(0.0535)	(0.0274)
SC	0.388***	0.269***	0.440***
	(0.0471)	(0.0698)	(0.0306)
ST	0.120**	0.0150	0.0814*
	(0.0551)	(0.0922)	(0.0468)
ASSETS	-0.0133***	-0.0213***	-0.0136***
	(0.00328)	(0.00476)	(0.00208)
INCOMEPC	2.0280***	1.6244***	1.8176***
	(0.9014)	(0.8419)	(0.5774)
HHEDUC	-0.0206***	-0.0267***	-0.0180***
	(0.00354)	(0.00527)	(0.00246)
Constant	-0.841***	-0.575***	-0.977***
	(0.0521)	(0.0760)	(0.0366)
Observations	10,823	4,798	26,398

Standard errors in parentheses
*** indicates significance at 1% ($p<0.01$), ** indicates significance at 5% ($p<0.05$),
* Significance at 10% ($p<0.1$)

Table 3 shows the estimates in the case of money lenders. We could see the marginal differences in the likelihood of obtaining the credit. Households belonging to OBC have better coefficients than the other two caste groups. Small and marginal farmers of OBC have coefficient values of 0.39 compared to SC (0.34) and ST (0.12). The same pattern is also seen in the case of medium 0.33 for OBC, 0.27 for SC and 0.44 for ST) and large farmers.

Households who belong to STs have difficulties compared to OBC and SC farmers in all categories. All the values of coefficients are significant except for medium ST farmers. Hence, we can say that there are caste differences in accessing credit. We can also notice the differences among the farmers within the class groups. Small marginal and large farmers are more likely to obtain credit than medium farmers. Assets don't matter for any farmers to obtain credit from moneylenders.

Table 4: Credit access of households

	(1)	(2)	(3)
Variables	**Banks**	**Cooperatives**	**Money lenders**
ASSETS	-0.00206	-0.00704***	0.00453**
	(0.00222)	(0.00217)	(0.00197)
INCOMEPC	2.2439*	1.4173***	1.7691
	(0.6273)	(0.6170)	(0.4183)
MEDIUM	-0.0796	-0.144***	-0.0265
	(0.0507)	(0.0482)	(0.0435)
SMALL MARGINAL	-0.0590*	0.0251	0.0122
	(0.0305)	(0.0297)	(0.0285)
Constant	-1.698***	-1.716***	-1.795***
	(0.0357)	(0.0349)	(0.0349)
Observations	32,930	37,382	35,746

Standard errors in parentheses
*** indicates significance at 1% ($p<0.01$), ** indicates significance at 5% ($p<0.05$), * Significance at 10% ($p<0.1$)

Table 4 gives the results regarding those who could get credit when they applied for a loan in the last five years. Table 4 shows the small-marginal and medium farmers' access to credit from banks, cooperatives, microfinance, and money lenders. Income plays a significant role in accessing credit when they apply for commercial banks compared to money lenders and cooperatives.

Conclusion

Using the IHDS 2 data and applying the Probit model to examine the differences in loan accessibility among the different social groups indicates that there are caste and class differences in accessing agricultural credit, i.e., small-marginal farmers and socially disadvantaged groups of farmers face difficulties accessing credit.There is non-inclusiveness in credit disbursement when we compare with farm scale size, with small farmers facing hindrances in getting loans, similarly socially disadvantaged groups facing hurdles in access to credit compared to other social groups.In contrast to the findings of(Asiedu, 2012; Kumar &Venkatachalam, 2018) studies, these findings indicate positive discrimination in accessing credit from commercial banks. The reasons for the differences in credit accessibility with the different lenders by different castes are yet to be ascertained. Do these differences indicate discrimination? Since this study tried to look at the demand side, probing the supply side i.e., examining loan repayment histories of the borrowers, would give the complete picture, which would help in taking affirmative actions and moving towards a more egalitarian society.

References

Asiedu, E., Freeman, J. A. and Nti-Addae, A. (2012). Access to credit by small businesses: How relevant are race, ethnicity, and gender? *American Economic Review,* 102 (1), 532–537.

Arrow, K. J. (1973). The theory of discrimination. In O. Ashenfelter & A. Rees (Eds.), *Discrimination in labor markets* (pp. 3–33). Princeton, NJ: Princeton University Press.

Altonji, J. G., & Blank, R. M. (1999). Chapter 48 race and gender in the labor market. In O. C. Ashenfelter & D. Card (Eds.), *Handbook of labor economics* (Vol. 3C, pp. 3143–3259). London: Elsevier.

Becker, G. S. (1971). The economics of discrimination. Chicago: University of Chicago Press

Burgess, R., R. Pande, and G. Wong. (2005). Banking for the poor: Evidence from India. *Journal of the European Economic Association,* 3 (2-3), 268–278.

Burgess, R., and Pande, R. (2005). Do rural banks matter? Evidence from the Indian social banking experiment.*The American Economic Review*, 95, 780–795.

Chaudhuriab K., Cherical Mary. (2012). Credit rationing in rural credit markets of India. *Applied Economics,* 44 (7): 803-812

Das, V., Das, E. V., Beteille, A., & Madan, T. N. (Eds.). (2003). *The Oxford India companion to sociology and social anthropology* (Vol. 1). Oxford University Press, USA.

Deshpande, A. (2000, May). Does Caste Still Define Disparity? A Look at Inequality in Kerala, India. *The American Economic Review,* 90 (2), 322–325.

Fafchamps, M. (2013). Credit constraints, collateral, and lending to the poor. *Journal of Development Economics*, 21(2), pp.79-100.

Ghurye, G. S. (1969). *Caste and race in India*. Popular Prakashan.

Kumar, S. M. (2013). Does Access to Formal Agricultural Credit Depend on Caste? *World Development,* 43, 315–328.

Kumar, S. M., & Venkatachalam, R. (2018). Caste and credit: A woeful tale? *The Journal of Development Studies*, 1–18.

Kumar, A., et al. (2017). Institutional versus non- institutional credit to agricultural households in India: evidence on impact from a national farmers' survey. *Economic Systems,* 41(3), 20-432.

Pallavi, C. (2007). Access to bank credit: Implications for Dalit rural households. *Economic and Political Weekly*, 42, 3219-3224.

Rajeev, M., et al. (2011). Credibility of equal access to credit.*Economic and Political Weekly,* 46(33), 76-79.

Mehrotra, N. (2011). Agriculture Credit: The Truth behind the Aggregate Numbers. *Economic and Political Weekly,* 46(42), 22-26.

Mohan, R. (2006). Agriculture Credit in India- status, issues and future agenda. *Economic and Political Weekly.*

Narayanan, S. (2016). The productivity of agricultural credit in India. *Agricultural Economics, 47*(4), 399-409.

Subbarao, D. (2012). Agricultural credit-accomplishments and challenges. *Speech delivered at the thirty years anniversary celebration of NABARD at Mumbai on July, 12.*

Thorat, S., & Attewell, P. (2007). The legacy of social exclusion: A correspondence study of job discrimination in India.*Economic and Political Weekly*, 42(1), 4141–4145.

Vijayabaskar, M., & Kalaiyarasan, A. (2014). Caste as social capital: The Tiruppur story. *Economic and Political Weekly*, 34-38.

18

Digital Financial Inclusion as a Gateway to Sustainable Development: An Exploration of the Informal Sector in Chennai

D. Sangavi and *Shobana Karthikeyan*

Introduction: Financial Inclusion and the Sustainable Development Goals (SDGs)

The world has become increasingly connected today than it was ever before. With more people having access to smart devices, the collective digital footprint of humanity has increased globally. However, the more pertinent issue to be addressed here is, whether the digital access necessarily translates into digital financial inclusion for the world's poor. Whether people at the bottom of the pyramid are advantaged by the digital world or do they remain excluded from this phenomenon is a question that needs to be answered at this critical juncture. The key question that countries face, especially against the backdrop of Covid-19 is, how can this digital revolution help nations reach the 2030 Sustainable Development Goals more quickly and effectively.

One of the catalytic yet realistic pathways for economic growth is through digital financial inclusion. Financial inclusion, by definition means *'affordable, effective, and safe financial services for everyone'*. Inclusive digital financial services could mean anything from mobile money, online accounts, electronic payments, insurance and credit and any other financial services that reach people who were formerly excluded (World Bank Compendium, 2018). It can be said that sustainable digital financial inclusion has the potential to drive and fuel the growth process of nations, particularly of the emerging economies.

Digital Financial Inclusion and Interlinkages with the Sustainable Development Goals (SDGs)

Digital financial inclusion provides low-income households access to tools that help increase their economic opportunities (SDG 1); provides farmers access to the funds needed to produce higher yields, thereby increasing agricultural productivity (SDG 2); helps household manage health emergencies better without being pushed into poverty (SDG 3); enables better financial management of education systems worldwide (SDG 4); empowers people, especially women to earn more and build assets, thereby fuelling greater gender equality (SDG 5); catalyses low-cost business models which create job opportunities world over and enables the growth of small businesses and eventual economic growth (SDG 8) (World Bank Compendium, 2018). Digital financial inclusion can bring the society towards reducing inequalities by giving low-income households new tools by which they could increase their incomes and improve their financial resilience (SDG 10). Even governments world over are better positioned to deliver services better using digital avenues than using cash. Therefore, it can be said that individuals, groups, communities, businesses and governments have the potential to perform and deliver better (SDG 11) in a more transparent and efficient digitally linked world. Therefore, access to digital financial services makes the journey towards Sustainable Development Goals (SDGs) easier for nations.

Digital Transactions: Meaning and Prevalence in India

Digital payments are transactions that take place via digital or online modes, with no physical exchange of money involved (www.razorpay.com). This means that both parties, the payer and the payee, use electronic mediums to exchange money (www.cashlessindia.gov.in). The digital financial service obtained from digital payment operators such Google pay, phone pay etc. Digital payment started flourishing in year 2016 though launch of 'Digital India Mission' encouraging Indian businessmen as well as others to adopt digital technology in order to strengthen the country's economy through paperless, faceless and cashless transactions (Singhal and Gupta, 2021). While the concept of digital payments had taken birth many years ago, it is presently trending at a faster pace, especially in the urban areas, after the Covid-19 pandemic, which imposed severe restrictions on the physical movement of people and goods worldwide.

The Covid-19 pandemic played an important role in revolutionising the digital financial services in India. India is steadily moving away from a cash-based economy and making great strides towards becoming a digital economy. Retailers started using digital payments system due to the convenience it offers. It is estimated that India's digital payments market will more than triple from $3 trillion to $10 trillion by 2026 (www.tele.net.in). Digital payments (non-cash) will constitute nearly 65 percent of all payments by 2026.

Digital Financial Inclusion and the Unincorporated Sector in India

In furthering our understanding of digital financial inclusion, it is important to look beyond the corporate sector and look at the unincorporated sector too. The unorganised sector in India employs 94 percent of the workers and produces 45 percent of the output and is widely prevalent both in the urban and the rural sectors. Street vending is an integral part of the urban informal economy, where the middle class and the urban poor access essential goods and services. The Government of India defines a street vendor as *'any person who is engaged in vending of articles, goods, wares, food items or merchandise of daily use or offering services to the public in a street, footpath, pavement and so forth from a temporary built structure or by moving from one place to another'* (Ministry of Housing & Urban Affairs). They are mostly rural migrants who come to cities looking for employment opportunities, and find that street vending gives them the independence of trade, and thus, a good livelihood opportunity (Kumar, 2022). Street vendors use cash for their day-to-day transactions and the street vending business is essentially propelled by cash transactions. But during and post Covid-19, the informal sector made great strides towards digital finance – towards accepting online payments from customers and paying their suppliers in the online mode.

While the street vendors have embraced digital payments in large numbers, it is worth investigating the effect of digitization of their payments on the volume of sales. Also, given the profile of the street vendors, the transition to digital payments could have been laden with challenges, which also merits a research response from the academia. Finally, it needs to be understood with the help of empirical research, if there has been any real economic empowerment of the street vendors on account of digital financial inclusion.

Objectives of the Study

The objectives of the study are as follows:

- To identify the perceptions about and reasons for accepting digital payments among street vendors along the Tambaram–Velachery Main Road in Chennai.
- To analyse the challenges faced by street vendors in adopting to digital payments.

Review of Literature and Research Gap

Digital financial inclusion has become an important topic of research interest for the academia especially after the Covid-19 pandemic. One of the key prerequisites in financial inclusion was the access to smart gadgets. This was investigated by Asongu et al. (2021). Few researchers were brought out theoretical frameworks that can be used by future researchers in their research activities. For example, Aziz and Naima (2021) build a framework for digital financial inclusion and Tram et al. (2021) look at a theoretical composite financial inclusion index of mobile money. Fraczek and Urbanek (2021) analyse digital payments from the angle of digital ticketing. Few other authors look at the role that trust plays in financial inclusion (Ghosh, 2021); and Xu (2020) research aspects of social trust and financial inclusion. The importance of digital financial inclusion in the eradication and reduction of the prevalence of poverty was brought out by Gutierrez- Romero and Ahamed (2021).

Digital inclusion studies have focused on several geographies. For example, Lutfi et al. (2021) analyse factors that led to the growth of Jordan Mobile Payment (JoMoPay). Huang et al. (2021) discuss trade openness and economic development in the European Union. Shen et al. (2021) perform a macro level study on the impact of digital financial inclusion in 105 countries worldwide. Bede Uzoma et al. (2021) discuss financial innovation and its impact on the eradication of hunger and poverty in the context of 27 Sub Saharan African nations. Wang and He (2021) research farmers' vulnerability in financial inclusion in China. Liu Yang and Youtang Zhang (2020), in an earlier study on China examine the impact of financial inclusion on the sustainable growth of small and micro enterprises. Bayero (2015) one of the early studies in this context, focuses on Nigeria. Soekarno et al. (2020) examine the influence of gender on financial inclusion in the context of Indonesia.

Focusing on the Indian context, there have been few key studies that have analysed the impact of digital financial inclusion. For example, Kanungo and Gupta (2021), discuss the impact of digital finance on overall social wellness in the Indian community. Pradhan et al. (2021) discuss the growth trends, financial inclusion initiatives, and ICT infrastructure development in about twenty Indian states. Undale et al. (2021), in the Indian context bring an important demographic dimension such as gender and research the comfort or security worry associated with the usage of an e-wallet. Amit Shankar and Biplab Datta (2018) employ a descriptive research design to test the adoption of Technology Acceptance Model (TAM) in India. Pal et al. (2018) in their research work, reveal the importance of digital payments and its acceptability among the people in India and suggest that the digital payments policy in India need to be further strengthened on account of widespread illiteracy among the users.

Moving over to the studies on digital financial inclusion with specific reference to street vendors or unorganised markets, studies are present, but are scant. Tanikan and Nittaya Wongtada (2018) study a technology acceptance model (TAM) in explaining technology adoption among street vendors in Thailand. Tiwari and Iyer (2017) work out a model for predicting vendors' decisions to switch to digital wallets in two Indian cities. Rizwana et al. (2021) study the factors that aid or thwart the adoption of e-wallets among street vendors in Bengaluru and Mysore. Falguni Mathews and Abhijit Bhosale (2021), in their research paper study the aspects that impact the use of digital payment systems by street vendors in Mumbai and Thane Region.

It is true that the academia has focused on the various aspects of digitalising businesses. Among the various studies attempted by the academia, studies on the impact of digital business on street vendors is scant. The impact of Covid-19, especially with the increased restrictions on people moving out of their homes and social distancing during has fallen majorly on the street vendors, who depend on regular footfalls for their businesses to thrive. It is time for the academia to focus on understanding the coping strategies that the street vendors had used to weather the negative impact of the pandemic. This study attempts to study the challenges faced by the street vendors in the specific context of digital transactions and their experiences with digital payments post-lockdown, particularly in the Chennai region. This is the gap in extant literature that this study seeks to address.

Methodology

This section aims at elaborating on the research methodology that was used in this research work. Primary data had been used to collect the data. To investigate the street vendors regarding usage of digital payments convenience sampling technique is being used. The total population of the study was the street vendors operating in the Tambaram - Velachery main road (both sides) around 300 vendors occupy the streets. A conveniently chosen sample of 74 street vendors who operated in the Tambaram-Velachery main road in Chennai were targeted for the purposes of the study. The study period covers the post lockdown period of January to April 2022. To approach street vendors regarding adoption of digital transactions, the researchers put together questions and had formulated a schedule.

Data Analysis and Interpretation

The demographic profile of the vendors can be seen in Table 1. Out of 74 respondents, the oldest was 67 years and the youngest was 17 years old. The average age of the street vendor is 36 years with the standard deviation of 10.912.

Table 1. Age of the respondents

	Age	Literacy level
N Valid	74	74
Mean	36.34	1.6622
Std. Deviation	10.912	1.01059
Minimum	17	0
Maximum	67	3

Table 2 shows the literacy levels of the respondents/street vendors. The respondents were given a choice to pick from No formal education, primary, high school, higher secondary, a degree course, higher than degree (such as a post graduate qualification) and a diploma/any other course. Completed level of education was taken into account. For example, if a respondent completed 7th standard or dropped out in the 9th standard, he/she was classified as having completed primary education only. In that sense, about 52.7 percent of the respondents had not completed high school education and the rest (47.2%) had education of high school and above. An interesting fact that emerged during the data collection was that while about 17.6 percent of the respondents did not have any formal education, they appeared well versed in running the business/using mobile phones/tracking payments. This is probably suggestive of why and how the usage of digital payments has proliferated in India.

Table 2: Literacy levels of the Street Vendors

No	Literacy levels (completed)	Frequency	Percentage
0	No formal education	13	17.6
1	Primary	22	29.7
2	High School	17	22.9
3	Higher Secondary	11	14.8
3	Degree	7	9.5
4	>Degree	0	0
5	Diploma/Others	4	5.5
	Total	**74**	**100.0**

Table 3 depicts the types of street vendors who were interviewed along the Tambaram-Velachery Main Road. An overwhelming 80 percent of the businesses were related to food and dealt with products of a perishable nature. 20 percent of the products belonged to the non-food-non-perishable category.

Table 3: Types of Street Vendors in the Sample

Items	Type of Product	Frequency	Percentage (%)
Food hawkers	Food/Perishable	8	10.8
Fresh juice	Food/Perishable	5	6.8
Fruits	Food/Perishable	16	21.6
Ice apple	Food/Perishable	3	4.1
Ice cream	Food/Perishable	2	2.7
Milk products	Food/Perishable	1	1.4

Items	Type of Product	Frequency	Percentage (%)
Sugarcane juice	Food/Perishable	8	10.8
Tea/Hot Beverage	Food/Perishable	5	6.8
Tender coconuts	Food/Perishable	3	4.1
Vegetables	Food/Perishable	5	6.8
Water Bottles	Food/Perishable	3	4.1
Baskets/Jute Bags	Non-perishable	4	5.4
Miscellaneous (Toys, knick-knacks, etc)	Non-perishable	11	14.9
Total		**74**	**100.0**

Figure 2 discusses the perceptions about digital payments and the reasons for accepting digital payments among street vendors in the Tambaram-Velachery Main Road in Chennai. Consumer preference can be seen as the driving force (82%) behind reasons for vendors adopting to digital payments. This is distantly followed by Covid-19 (12%) acting as a catalyst and finally followed by the practical convenience of not having to travel to a bank (6%) and ATM ranks last among the perceptions. A lot of changes in the businesses especially on the digitalization are generally attributable to the restrictions imposed on account of the Covid-19 pandemic. The same is reflected in this study in a round-about way. Covid-19 has changed consumer preferences in adapting to digital transactions leading to consumers demanding the option of digital mode of payments from the vendors, thereby forcing the vendors to provide that facility for their consumers. So, in a way, it is possible to conclude that while Covid-19 has directly impacted the perceptions of vendors in a smaller way, it has had a huge indirect impact in an indirect way via changing the perceptions and expectations of the consumers.

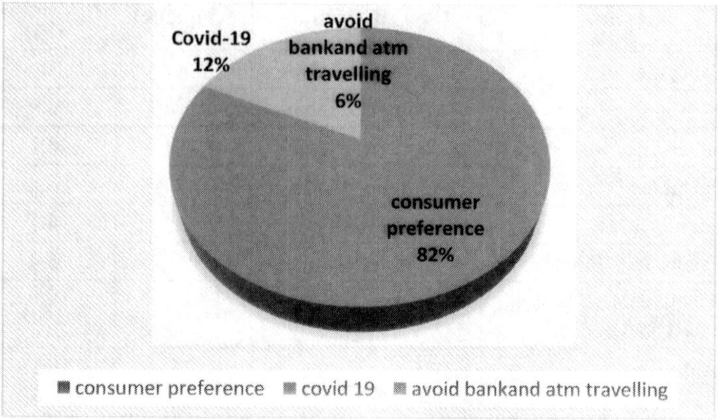

Fig. 1. Reasons for adapting to digital payments.

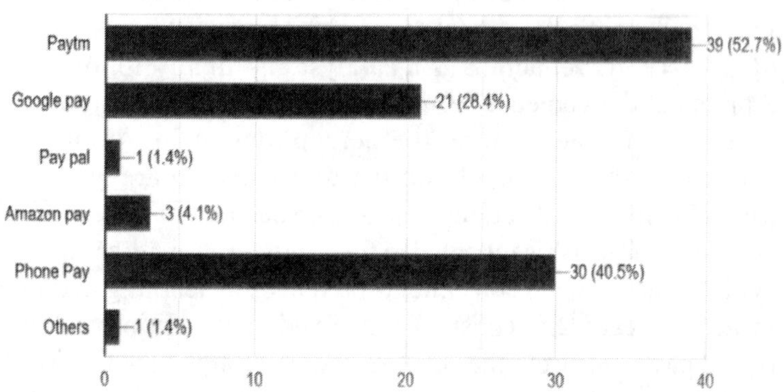

Fig. 2: Payment platforms used in receiving digital payments.

Figure 2 describes the popularity of the payment platforms among the vendors in receiving digital payments. The vendors were provided a checklist and were asked to tick the payment platforms they used. In many instances, the vendors subscribed to more than one service. It can be seen in Figure 2 that Paytm and Phone Pay are the most used platforms among the vendors at 52.7 percent and 40.5 percent respectively. Google Pay is used by about 28.4 percent. Paypal, Amazon Pay and Others are used by very negligible number of vendors. When the researchers looked deeper into the data, the vendors who used Pay pal, Amazon Pay and either also used PayTm or Phone Pay to manage their regular foot falls.

Figure 3 explains the perceptions of vendors regarding usage of digital payments. A point to note is that these perceptions are different from perceptions regarding adoption of digital payments. Three factors stood out when asked about vendors' perceptions of usage of digital payments. 1. Ease of receiving payments. 2. Avoids time consumption and wastage, meaning it is quick and efficient. 3. It avoids the need for travel to a bank or to an ATM to deposit/withdraw cash.

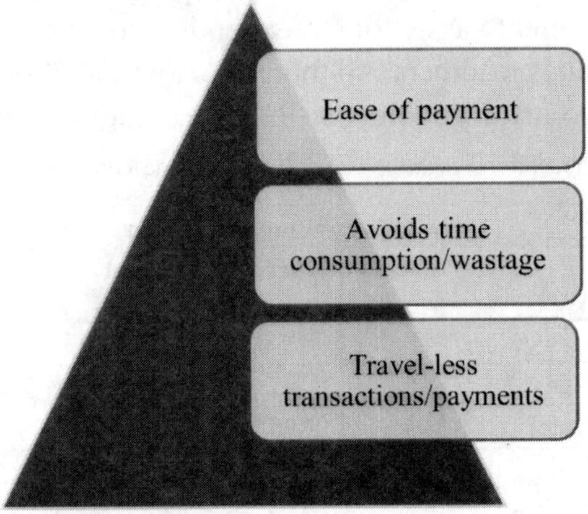

Fig. 3: Perceptions regarding digital payments.

Table 4 shows that perception of 91 percent street vendors, who agreed for the "ease of payments" and remaining eight percent have disagreed. The reasons for disagreement of the vendors regarding ease of payments are discussed in the next session where they put forth certain issues such as server troubles and the chance of being defrauded in online transactions.

Table 4. Perceptions regarding ease of payment

		Frequency	**Percent**	**Valid Percent**	**Cumulative Percent**
Valid	Disagree	6	8.0	8.1	8.1
	Agree	68	90.7	91.9	100.0
	Total	**74**	**98.7**	**100.0**	

Results from Table 5 show that 94.6 percent of the vendors agreed that online payments reduce time consumption in settling transactions and time wastage. In the case of cash transactions, there are many situations when customers are unable to pay the exact amount for the products purchased, so it is a permanent hassle for street vendors to give the balance to demanding customers. All this costs time and effort, and this can be conveniently avoided in the case of digital transactions.

Table 5. Perceptions regarding avoidance of time consumption/wastage

Avoid time consumption/wastage					
		Frequency	**Percent**	**Valid Percent**	**Cumulative Percent**
Valid	Disagree	4	5.3	5.4	5.4
	Agree	70	93.3	94.6	100.0
	Total	**74**	**98.7**	**100.0**	

Another important perception in the usage of online payments is the perception of avoidance of travel to and from banks. Substantial time can be lost in travel to financial institutions and that can be avoided when payments are made online. It can be observed from the data (from Table 6) that almost two-thirds of the vendors concur with that opinion and the rest do not. The reason for disagreement for "Travel-less payment" mentioned by street vendors is that they had to frequently travel to bank for cross checking the payment transfers in their account.

Table 6: Perceptions regarding avoidance of travel or travel-less payments

Travel less payment					
		Frequency	**Percent**	**Valid Percent**	**Cumulative Percent**
Valid	Disagree	49	65.3	66.2	66.2
	Agree	25	33.3	33.8	100.0
	Total	74	98.7	100.0	

Difficulties Associated with Digital Payments:

As the researchers analysed the positives of digital finance, it was considered equally important to understand if there were any difficulties associated with online payments. Two problems were considered important impediments. 1. Server problems and 2. Afraid of being betrayed or cheated (the case of online fraud).

From Table 7, it can be observed that around 62.2 percent of the vendors faced server problems. During personal interviews, the vendors also said that few transactions would remain incomplete because of server issues, and hence the customer would leave without completing the transaction, sometimes, by paying cash, and sometimes, not paying too.

Server Problems

Table 7: Challenges faced in adopting digital payments – Server problems

Server Problems					
		Frequency	**Percent**	**Valid Percent**	**Cumulative Percent**
Valid	Agree	46	62.2	62.2	62.2
	Disagree	28	37.8	37.8	100.0
	Total	**74**	**100.0**	**100.0**	

Table 8 shows details of fear of being betrayed. During the Covid-19 outbreak, the society has witnessed many changes. New work-level changes like changes like working from home, using online services and online vendors, and becoming more active virtually, etc pose considerable cybersecurity challenges. Cheating in online transactions also was/is a common phenomenon. This study also seeks to understand if fear of being betrayed was prevalent in the minds of the online vendors. Less than half of the vendors (44.6%) of the people were scared of being betrayed/duped in online transactions and around 55 percent of the vendors trusted the system such that they did not think they would be duped online.

Afraid of being betrayed

Table 8: Challenges faced in adopting digital payments – Fear of being cheated.

Afraid of being betrayed due to less Knowledge about digital transactions.					
		Frequency	**Percent**	**Valid Percent**	**Cumulative Percent**
Valid	Agree	32	43.2	43.2	43.2
	Disagree	42	56.8	56.8	100.0
	Total	**74**	**100.0**	**100.0**	

Table 9: Correlation between education and Fear of online duping

Correlations			
		Afraid of being betrayed due to less Knowledge about digital transactions.	**Education**
Afraid of being betrayed due to lack of Knowledge about digital transactions.	Pearson Correlation	1	.576**
	Sig. (2-tailed)		0.000
	N	74	74
Education	Pearson Correlation	.576**	1
	Sig. (2-tailed)	0.000	
	N	74	74

Table 9 seeks to understand the correlation between education and fear of being betrayed due to less knowledge about digital transactions. This speaks about the risk averseness of vendors in this context. A significant positive correlation of .576 is observed between the two variables in this study. This explains that as the education levels increase, the knowledge/exposure regarding digital cheating is prevalent. And those with lower levels of education are blissfully unaware of the dangers of digital transactions.

Table 10 discusses the perceptions of vendors regarding incomes and their adoption of digital payments. The researchers asked the question whether they perceived an increase in income after adopting digital payments. Around 34 percent of the respondents did not provide any definitive answer to the question. Around 5 percent of the respondents answered in the negative. And a decent majority of the respondents (around 66% of them) agreed in the affirmative saying that adopting digital payments led to an increase in their income, as per their perception.

Table 10: Perceptions of the vendors regarding increase in income after adopting digital payments

Perception of the question: Has your income increased after adopting digital payments?	Frequency	Percentage (%)
Strongly disagree	1	1.4
Somewhat Disagree	3	4.1
Neutral – Neither agree nor disagree – Can't Say	25	33.8
Somewhat Agree	20	27.0
Strongly agree	29	39.2
Total	**74**	**100.0**

Conclusion

The vendors operating in the unorganised are an important entity, especially in urban India, where they make their presence felt along the streets, street corners, pavements, bus stands, railway stations, besides others. They have their regular set of customers who visit them and they are their steady source of income/ livelihood. The unorganised sector fills an important gap that the organised sector does not fill. It provides the set of goods, which the organised sector does not offer, or even if it offers, does not do so as efficiently as the unorganised sector. E.g., Vending of flowers. That provides the space for the unorganised sector to flourish in the urban areas.

This study aims to find out the impact of digital transformation in trade transactions among the street vendors in Tambaram-Velachery Main Road in Chennai. According to National Association of Street Vendors of India (2013), the occupational health and safety of the street vendors and their conditions of work are often a serious concern and welfare and developmental measures do not reach them. The overall findings of the study reveal the reason for adopting digital payments among street is mainly consumer preference. However, the unorganised sector still prefers to deal in cash because of 'scare of being betrayed due to less knowledge about operating of digital payments. Moreover, this mode of payment makes the vendors to visit bank frequently which affects their day-to-day business.

This study found that the inclusion of digital payments system among the street vendors increases the income, investment and the volume of sales. It leads to enhancing their economic empowerment in the society. It is important to observe that the street vendors showed a high level of resilience and switched to digital transactions. The vendors were able to access services which they did not have access to. Digital transactions, when looked at with some depth, speaks of a deep-rooted change that has taken place in the Indian society. From being an almost all-cash economy, India has come a long way. Digital financial inclusion has brought about changing habits and attitudes among the people (both on the vendors' side and the customers' side), the clamour for better infrastructure, cyber literacy and informed customers, besides others. All of these are important preconditions for the attainment of the 2030 sustainable development goals (SDGs).

References

Abhijit Bhosale, (2021). "Impact of Covid-19 on Digital Payment Transactions in India". *Shodh Sanchar Bulletin,* Vol. 11, No. 41, 1-20.

Asongu, S. A., Biekpe, N., & Cassimon, D. (2021). On the diffusion of mobile phone innovations for financial inclusion. *Technology in Society, 65*, 101542.

Aziz, A., & Naima, U. (2021). Rethinking digital financial inclusion: Evidence from Bangladesh. *Technology in Society, 64*, 101509.

Bayero, M. A. (2015). Effects of Cashless Economy Policy on financial inclusion in Nigeria: An exploratory study. *Procedia-Social and Behavioral Sciences, 172*, 49-56.

Bede Uzoma, A., Omankhanlen, A. E., Obindah, G., Arewa, A., & Okoye, L. U. (2020). Digital finance as a mechanism for extending the boundaries of financial inclusion in sub-Saharan Africa: A general methods of moments approach. *Cogent Arts & Humanities, 7*(1), 1788293.

Chopra, R., (2017). Impact of Demonetization on Indian Economy, *Discover @ Georgia Southern*, EBSCO Publishing Service Selection Page, 8 June.

Dhanalakshmi, S., Komalavalli, K. and Hemalatha, R., (2019). The Cashless Economy and its Effects among the Salaried Employees in Chennai City, *International Journal of Recent Technology and Engineering*, 8, 4S3, 1-13.

Frączek, B., & Urbanek, A. (2021). Financial inclusion as an important factor influencing digital payments in passenger transport: A case study of EU countries. *Research in Transportation Business & Management, 41*, 100691.

Franciska, A. M., & Sahayaselvi, S. (2017). An overview on digital payments. *International Journal of Research, 4*(13), 2101-2111.

Ghosh, S. (2021). How important is trust in driving financial inclusion? *Journal of Behavioral and Experimental Finance, 30*, 100510.

Gutiérrez-Romero, R., & Ahamed, M. (2021). Covid-19 response needs to broaden financial inclusion to curb the rise in poverty. *World Development, 138*, 105229.

Hasanul Banna, (2020). The role of digital financial inclusion on promoting sustainable economic growth through banking stability: Evidence from Bangladesh. *Research Gate*, Volume 29 (ISSN: 1607-8373).

Huang, R., Kale, S., Paramati, S. R., & Taghizadeh-Hesary, F. (2021). The nexus between financial inclusion and economic development: Comparison of old and new EU member countries. *Economic Analysis and Policy*, *69*, 1-15.

Indira, D., (2014). A Study on the Organising of Street Hawking Business, *International Journal of Management and Commerce Innovations,* 2, 1, pp.280-288.

Kaka, N., Madgavkar, A., Kshir, S., Gupta, R., Manyika, J., Bhal, K. and Gupta, S., (2019). *Digital India: Technology to Transform a Connected Nation*, Mckinsey Global Institute. pp.1-16.

Kanungo, R. P., & Gupta, S. (2021). Financial inclusion through digitalization of services for well-being. *Technological Forecasting and Social Change*, *167*, 120721.

Khalil, M. N., *(2007). The influence of trust on Internet banking. Journal of Internet Banking and Commerce*, Vol. 12, No. 2, pp. 1-10, ISSN 1204-5357.

Kumar Spm. (2022). A Study on the impact of Cashless Transactions among Street Vendors. International Journal of Early Childhood Special Education. 14. 3471-3475.

Kumar, S. (2022). A Study on the impact of Cashless Transactions among Street Vendors. *International Journal of Early Childhood Special Education,* Vol.14. https://doi.org/3471-3475. 10.9756/INT-JECSE/V14I1.420.

Lutfi, A., Al-Okaily, M., Alshirah, M. H., Alshira'h, A. F., Abutaber, T. A., & Almarashdah, M. A. (2021). Digital financial inclusion sustainability in Jordanian context. *Sustainability*, *13*(11), 6312.

M. Rizwana & Padmalini Singh & P. V. Raveendra, 2021. "Promoting Financial Inclusion Through Digital Wallets: An Empirical Study with Street Vendors," Springer Books, in: Ananda S. & Dharmendra Singh (ed.), *Financial Inclusion in Emerging Markets*, pp. 281-293, Springer.

Mathews, F., & Bhosale, A. (2021). Understanding the Factors Influencing Street Vendors towards Use of Digital Payment Systems in India. In *2021 9th International Conference on Reliability, Infocom Technologies and Optimization (Trends and Future Directions) (ICRITO)* (pp. 1-7). IEEE.

Ministry of Housing and Urban Affairs, (2019). Handbook of Urban Statistics, Government of India. 2019.

Nandru, P., Chendragiri, M., & Velayutham, A. (2021). Examining the influence of financial inclusion on financial well-being of marginalised street vendors. *International Journal of Social Economics,* (Issue publication date: 13 July 2021), 1-24. ISSN: 0306-8293

Nariyanuri, S.S., (2020). India Mobile Payments Market Report, *S & P Market Intelligence*, https://www.spglobal.com/marketintelligence/en/documents/ indiamobilepayments_2020finalreport.pdf.

Norm & Pearson, (2018), The influence of trust on Internet banking acceptance. *Journal of Internet Banking and Commerce,* Vol.no. 12 (ISSN 1204-5357.), pp.1-10.

O'Sullivan, K., Clark, S., & Marshall (2021). A Just Digital framework to ensure equitable achievement of the Sustainable Development Goals. *Nature Communications*.

Pal, J., Chandra, P., Kameshwaran, V., Parameshwar, A., Joshi, S., and Johri, A., (2018). Digital Payment and Its Discontents: Street Shops and the Indian Government's Push for Cashless Transactions, *Proceedings of the 2018 CHI Conference on Human Factors in Computing Systems*, Paper No. 229, 1–13.

Pipitwanichakarn, T., & Wongtada, N. (2018). Mobile commerce adoption among the bottom of the pyramid: A case of street vendors in Thailand. *Journal of Science and Technology Policy Management*.

Pradhan, R. P., Arvin, M. B., Nair, M. S., Hall, J. H., & Bennett, S. E. (2021). Sustainable economic development in India: The dynamics between financial inclusion, ICT development, and economic growth. *Technological Forecasting and Social Change, 169*, 120758.

Punia, V., Khera, A. and Sehrawat, J.S., (2018). A Critical Analysis of Demonetization in India: Its Success and Failure, *International Journal of Research in Advent Technology*, Vol. No. 6, 5, 833-839.

Rajanna, K.A., (2018). Perceptions and Awareness of Customer towards Cashless Transaction: A Study, *International Journal of Application or Innovation in Engineering and Management,* Vol no. 7, 113-116.

Rouibah. K, (2015). *"Electronic payment systems use and satisfaction in an Arabic Country: Evidence from Kuwait", Issues in Information Systems*. vol. 16, no. II, pp.149-160.

Shen, Y., Hu, W., & Hueng, C. J. (2021). Digital financial inclusion and economic growth: a cross-country study. *Procedia Computer Science, 187*, 218-223.

Singh, S., & Rana, R. (2017). Study Of Consumer Perception of Digital Payment Mode a Research. *Journal of Internet Banking and Commerce,* Vol. 22, 1-18.

Singhal, Rashi and Gupta, Abhilasha, (2021). Impact of Covid-19 on Digital Payment Services at Towns and Villages. *International Journal of Creative Research Thoughts* (IJCRT) 2021, Available at SSRN: https://ssrn.com/abstract=3947798

Sivasubramanian. K and Rajendran. G, (2020). Evaluating the Impact of Digital Transformation on Economic Conditions of Unorganised Small and Petty Traders in Bengaluru . *International Journal of Economics, Commerce and Research,* Vol.10, 53-60.

Soekarno, S., Mambea, I. Y., & Setiawati, M. (2020). Women and Digital Financial Inclusion in Indonesia as Emerging Market. *International Review of Management and Marketing, 10*(5), 46.

Iyer, L. S. (2018). Adoption of digital wallets by petty vendors post demonetization in India: a prediction approach. *Asian Journal of Research in Social Sciences and Humanities, 8*(6), 117-130.

Tram, T. X. H., Lai, T. D., & Nguyen, T. T. H. (2021). Constructing a composite financial inclusion index for developing economies. *The Quarterly Review of Economics and Finance.*

Undale, S., Kulkarni, A., & Patil, H. (2020). Perceived eWallet security: impact of Covid-19 pandemic. *Vilakshan-XIMB Journal of Management.*

UNSGSA *Igniting SDG Progress through Digital Financial Inclusion* (2018) Available Online: DG_Compendium_Digital_Financial_Inclusion_September_2018.pdf

Wang, X., & He, G. (2020). Digital financial inclusion and farmers' vulnerability to poverty: Evidence from rural China. *Sustainability, 12*(4), 1668.

World Bank Group. (2013). *Global financial development report* 2014: Financial inclusion (Vol. 2): World Bank Publications.

Xu, X. (2020). Trust and financial inclusion: A cross-country study. *Finance Research Letters, 35*, 101310.

Ying, Tay.L, Toong, Tai, H., & Siang Tan, G. (2022). Digital financial inclusion: A gateway to sustainable development paper. *Journal of International Economics, Volume 8*(6 June 2022). https://doi.org/, e09766: https://doi.org/10.1016/j.heliyon.2022.e09766

https://razorpay.com/learn/digital-payments-india-definition-methods-importance/

https://tele.net.in/going-cashless-innovations-pave-the-way-for-digital-payments-in-india/https://thewire.in/economy/indias-unorganised-sector-is-being-engulfed-further-marginalised.

Index

B

C

F

G

H

R

S